The Poetries of Hotels
诗意酒店

深圳市创扬文化传播有限公司 策划　徐宾宾 编

凤凰出版传媒集团 | 凤凰联动・空间
江苏人民出版社 | IFENGSPACE

图书在版编目（CIP）数据

诗意酒店 / 徐宾宾编. 一南京：江苏人民出版社，2011.3
ISBN 978-7-214-06773-9
Ⅰ. ①诗… Ⅱ. ①徐… Ⅲ. ①酒店－室内设计－世界－图集
Ⅳ. ①TU247.4-64

中国版本图书馆CIP数据核字(2011)第002376号

诗意酒店 徐宾宾 编

责任编辑：刘 焱
出　　版：江苏人民出版社
发　　行：天津凤凰空间文化传媒有限公司
销售电话：022-87893668
网　　址：http://www.ifengspace.cn
集团地址：凤凰出版传媒集团（南京湖南路1号凤凰广场A楼 邮编：210009）
经　　销：全国新华书店
印　　刷：利丰雅高印刷（深圳）有限公司
开　　本：878毫米×1160毫米 1/12
印　　张：24
字　　数：145千字
版　　次：2011年3月第1版
印　　次：2011年3月第1次印刷
书　　号：ISBN 978-7-214-06773-9
定　　价：328.00元（USD 66.00）
（本书若有印装质量问题，请向出版社发行部调换）

序

Preface

梁景华

香港十大设计师、PAL设计事务所有限公司创办人及首席设计师

酒店设计 Hotel Design

For years, among a variety of interior designs, hotel design is the large project I have paid the most attention. Every hotel has something of its own, whatever the brand positioning, pattern of operation or historical legend, they mean much from the angle of hotel design. Whether it's implied or expressed, we can get some ideas from the design to express the hotel's unique features. The hotel design we take part in is an integrated, all in one discipline, from the conceptual design to deep design until to the treatment of soft furnishings, and such an integrated design is the most ideal.

Personally, I lay an emphasis on using natural materials as nature is most beautiful and natural materials can express their own strong points most, for example, I was ever invited to act as a guest lecturer on a world forum, during which I presented the merits of "wood" and it turns out such a material has 6, 7 virtues, one of which is its textures, grains being totally simple and natural, unlike porcelain or plastic products, you won't find the identical woods. Apart from a unique appearance, it has a fresh fragrance, also feels warm, with its existence, the space would immediately come back to life. Another material is a rock with a stronger creativity, capable of being polished into a glossy surface or left a coarse texture without any polishing, both being able to increase the space's glamour and strongly communicate a naturally grand feel. Recently, I have even taken advantage of some cheap materials that seem ordinary and changed them from the decayed to the miracle, which were immediately transformed into a luxurious decorative wall. What are they? The answer is coconut shells; nobody knows the wall is covered with such material even after it was finished. Therefore, when looking at a thing, we should never see only one side, instead, we need to work hard to discover other interesting sides on our own.

Moreover, I hope to reflect the local special cultures. The resort hotel in Sanya is shown in the form of a beautiful flower-strelitzia reginae so as to praise the local enchanting paradise, exuding a cozy, leisurely air. Through unlimited inspirations, the space becomes more interesting.

多年来，在各种室内设计范畴之中，酒店设计是最令我专注的大型项目。每一间酒店都有其独特之处，不论是酒店本身的品牌定位，还是运作方式，或是历史传奇，对酒店设计的一环来说，都是极具意义性的。无论是含蓄或是明显，我们都可以从设计里得到一些灵感，以此表达出酒店独一无二的特色。我们参与的酒店设计是整合性的，由概念设计到深化设计，以至软装处理都是一条龙的专业，这样的整体设计是最理想的。

我个人着重天然物料的运用。自然是最美的，最能表达物料本身的优点。我曾经被邀请担任一个世界论坛的演讲嘉宾，我把“木”的优点呈现出来，总共有六七点，它的纹理是浑然天成，不像陶瓷或塑胶制成品，是没有完全相同的。除了外观独特，它的香气是清新的，而且带来温暖的感觉，空间立时显得生机勃勃。另一种素材，就是岩石，它的创造性更强，可以经打磨成光滑的层面，或是不经琢磨的粗糙质感，两者都能增添空间的魅力，强烈表达出一种自然宏伟的感觉。最近，我更利用一些看似平平无奇的便宜物料，在其中一个酒店项目里，将“它”化腐朽为神奇，顿时变成豪华的装饰墙面，这种物料是什么？就是椰子壳，设计完成后谁都不知道就是运用此物料做成的。所以当看待事物时，不要只顾看一面，你需要自己去用心发掘另一面的有趣点。

另外，我希望可以将当地特有的本土文化表达出来。位于三亚的度假酒店，就是以一种美丽的花卉和天堂鸟花的形态，赞颂当地的迷人天堂，散发一缕写意优闲的气息。通过无限的灵感，空间变得更有意思。

Contents

目录

享受生活，彰显自我

Enjoy Life, Express Oneself

i-SUITE酒店

i-SUITE Hotel

设 计 师：Simone Micheli

设计公司：Simone Micheli Architectural Hero

项目地点：意大利里米尼

项目面积：5500平方米

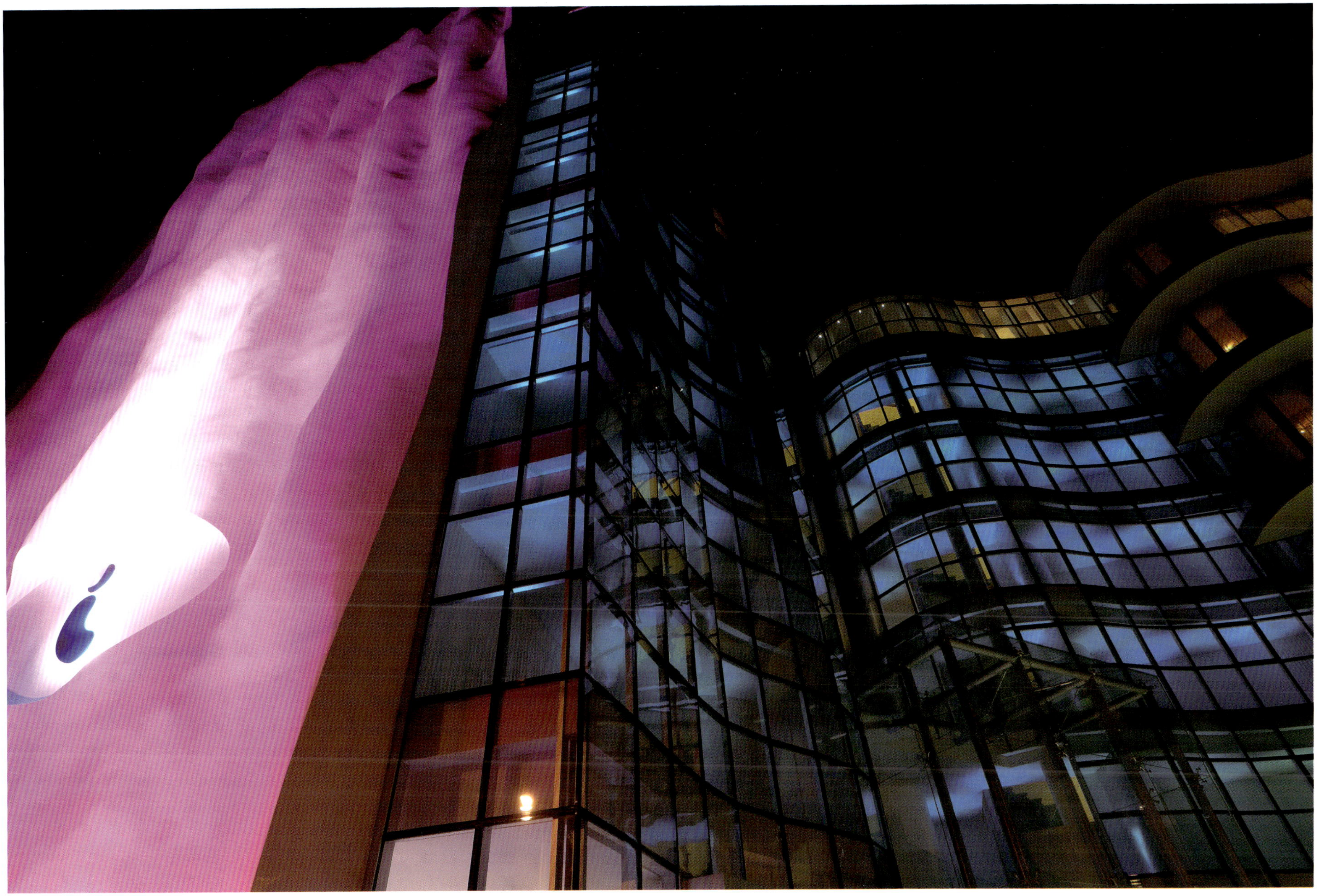

I-SUITE takes its inspiraton from the changes, from the thrust of a light ray fltering and thorough views from the glass. The furnishing is not rectlinear. There is a contrast between static state and dynamic state. It epitomizes the city of Rimini which does not make any stops and does not want to sleep. I aimed at shedding a new light on the hotel sector. We designed the plan that underlines lightness and purity by using glass. We used soft colors to conciliate the sunshine, while white rippled walls like sinuous dresses beckoning the guest to follow them.

I aim at making a main focus that stirs up senses by means of shapes, images, colors, lights, materials and by creating a scenario hovering among transcendence and immanence, concreteness and abstraction, dream and reality. Nowadays living in a luxurious space means being able to love and being loved in a genuine and absolute way. It means being surrounded by simplicity and having our attention drawn by a grass blade waving in the wind. It means regaining the truth of life, living by appreciating the beauty of each day. It means reconsidering, dreaming, starting to love again without asking anything back, and giving without having anything back. It means building without destroying, and creatng processes with a strong ethical sense and generators of healthy beauty.

水般的柔滑
像少女的裙摆
在风中起舞

I-SUITE的灵感来源于变化、过滤光线以及从玻璃里观察清晰外景的想法。家具都是非直线型的。空间动静相结合，俨然一座永不歇业的里米尼不夜城的缩影。设计的本意是希望赋予酒店一些新意。设计师用玻璃突出了空间的光亮的纯净，并用柔和的颜色调和刺眼的阳光。洁白的波浪形墙面像起伏的衣裙一样，吸引客人跟随它畅游其中。

设计师希望利用造型、形象、色彩、灯光和材质制造一个可以刺激感官的焦点，营造一种环境，使人徜徉于内蕴、具体与抽象、梦境与现实之间。今天，奢侈空间中生活形态应该被定义为可以感受真切的爱与被爱。一边置身于简洁的空间，一边欣赏草叶随风摆动的舞姿，每天都能享受美景。它也意味着反思、梦想、重新开始不求回报的爱。它标志着建立而非破坏，它是本着强烈的责任感而进行的创造活动，是健康美丽的起源。

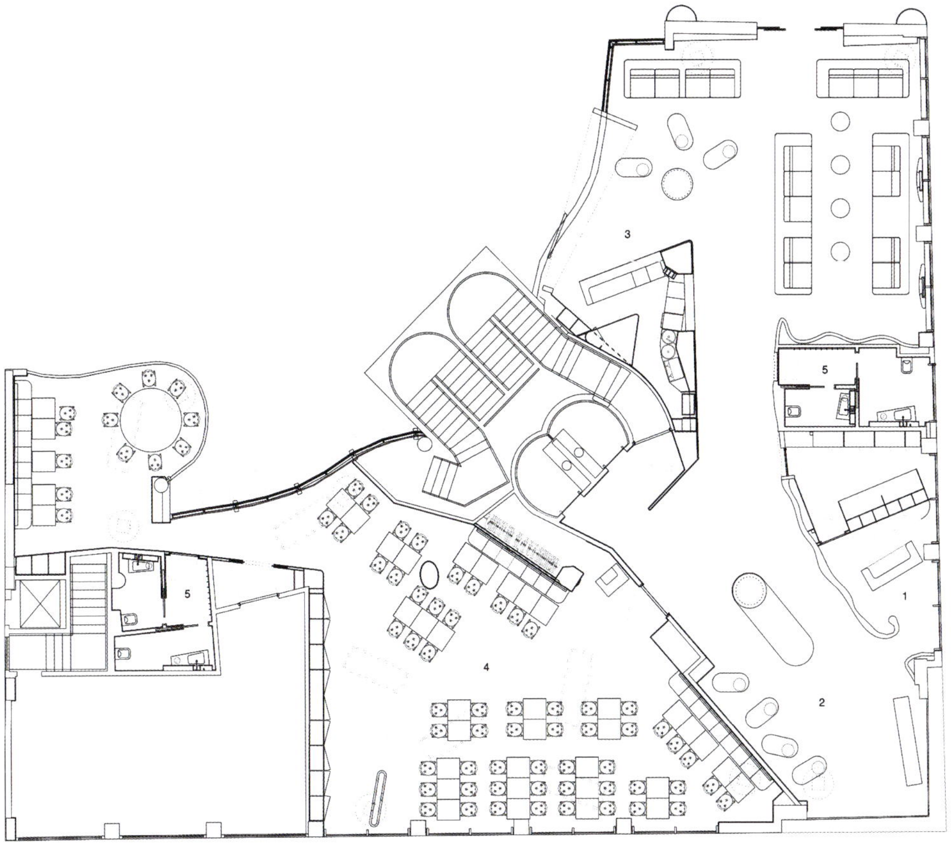

令人瞩目的建筑
An Emphatic Building

Alt酒店
Alt Hotel

设计公司：LEMAYMICHAUD建筑设计公司
项目地点：加拿大魁北克
项目面积：9500平方米
项目摄影：Pierre Bélanger

The design of a hotel within a recently developed commercial, cultural and corporate quarter of suburban Montreal represented an interesting architectural challenge and the opportunity to introduce a durable and significant presence in a constantly evolving environment.

The exterior treatment of the building is refined, sophisticated and was developed with a sense of design continuity. It is sophisticated in form, proportion and arrangement of its windows and choice of colours. It's continuity expressed in the choice of prefabricated concrete for the exterior cladding. Featured on the front and back facades are two suspended gardens affording large views over the nearby Monteregian Hills. The lighting of the facades provides a signal in the night without resorting to commercial signage.

In the interior, the concrete structure is largely exposed: columns in the public spaces, exposed concrete ceilings in the guestrooms and conference centre. Natural light filters abundantly through generous windows particularly in the rooms, where louvers also allow guests to control the intake of ventilation.

The typical room innovates with its unique arrangement and the use of glass and other non-traditional finishes. Warm coloured natural woods and plastic laminates are combined providing solidity and ease of maintenance. The generous public spaces are luminous, filled with stimulating colours and finishes. The lighting schemes of these spaces were as carefully designed to enhance the communication between the ground and second floors.

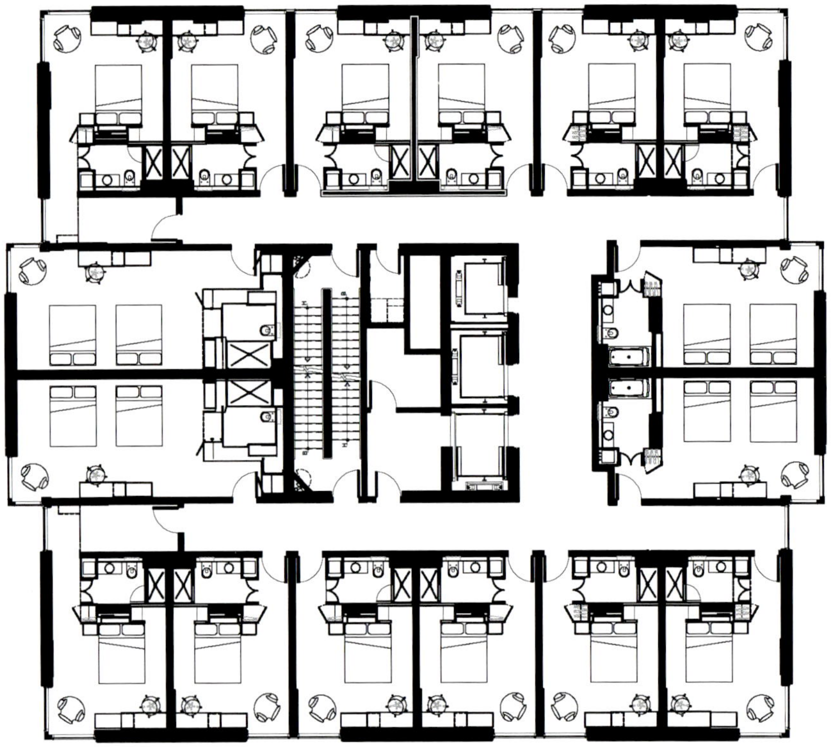

这家酒店位于蒙特利尔市郊新近开发的一片商业、文化和办公区，其完成了一次有趣的建筑挑战和对良机的把握：在持续演变的环境里引入一个经久耐用、令人瞩目的建筑实体。

酒店的外部处理精致、复杂，具有设计连续感。它的复杂体现在形式、比例、窗户的安排及颜色的选择方面。它的连续性表现在选用预制混凝土作为外墙覆层。建筑的前后外墙有两个悬空花园，在这里能看到蒙特里根丘陵的广阔风景。外墙的照明在晚上形成一个显著标志，从而无需借助商业标志。

酒店内部，混凝土结构大多裸露，如公共空间里的柱子、客房和会议中心里的混凝土天花。充足的自然光线透过宽大的窗户涌入房内，房间里的百叶窗也能让客人按自身需求控制通风。

标准房因为独特的布置，使玻璃的使用及其他非传统的修饰显得新颖别致。暖色系天然木材和层压塑料相结合，家具显得既坚固又易于打理。宽敞明亮的公共空间充满了能够刺激感官的颜色和修饰。这些空间的照明经过精心设计，提升了底层与第二层之间的沟通。

精致的设计
让建筑的魅力
令人瞩目而难忘

按博物馆规格修建的城市酒店
A City Hotel Following Museum Specifications

Met酒店
The Met Hotel

设 计 师：Zeppos-Georgiadis & Associates
项目地点：希腊塞萨洛尼基
项目摄影：Vaggelis Paterakis

The Met Hotel, conveniently situated on the harbour of Thessaloniki, consists of total 212 guestrooms and suites affording great city and port views. It also houses 13 meeting rooms accommodating up to 750 people, gourmet restaurants, atmospheric bars, a fabulous spa, a comfortable VIP lounge and other facilities and services that ensure an unforgettable stay. Two restaurants, 'Chan' designed by Andy Martin and Avenue 48 provide Pan-Asian and contemporary cuisine. Besides, the Long Bar and the Sky Bar on its roof garden feature a cosmopolitan atmosphere, panoramic views and hip music. In the Spa, the design language is kept simplistic and edgy, where exist indoor heated swimming pool with hydrotherapy, Jacuzzi, steam rooms, solarium, hair salon, fully equipped treatment and massage rooms.

There is a strictness and seriousness expressed in the interiors which is put forward through the straight edge clear cut furniture combined with artistic accessories and dynamic artwork. Every interior has the same design approach, maximizing and minimizing the levels of intensity according to its function. In the corridors leading to the bedrooms we experience volume and warmth through the structural wood panel forms. Arriving inside the bedrooms, there is the presence of wood, lacquer and glazing. Three finishes combined with smooth fabrics to create a luxurious but approachable space. Just looking around the lobby, you will only recognize items and furniture from high enders.

The Met Hotel is a leading example to contemporary design. It has form, ambience, luxury, sexiness and comfort. A building that understands the language of the visitor and the essence of a very important word in the Hotel Industry 'exclusive design'.

360° 城市的风景
曼妙多姿
新潮的气息
独一无二的设计

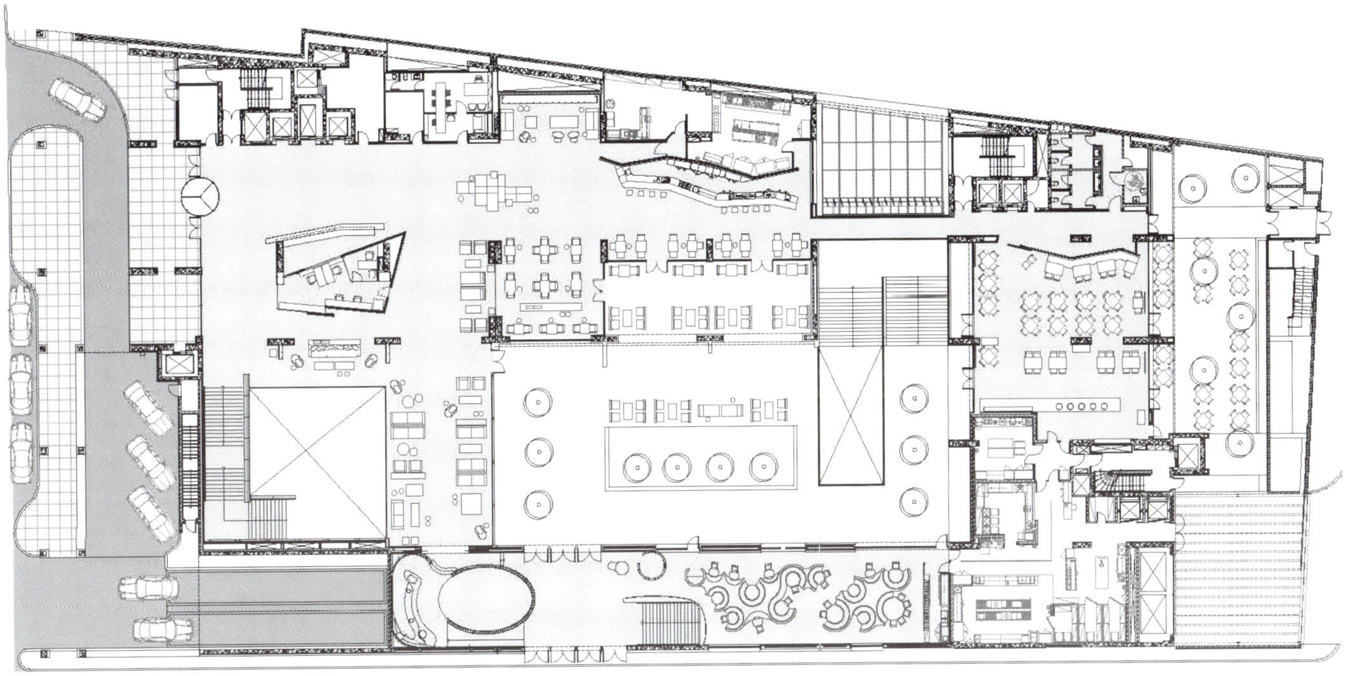

位于塞萨洛尼基港口的MET酒店交通位置便利，包括212间可以观赏到美妙城市风景和港口风景的客房和套房、能接待750人的13间会议室、美食餐厅、氛围酒吧、绝佳的水疗中心、舒适的贵宾休闲吧及其他设施和服务，确保留给客人难以忘怀的入住体验。两间餐厅——Andy Martin设计的“Chan”餐厅及另外一间餐厅Avenue（大道）48，供应亚洲当代新菜色及国际化佳肴。另外，屋顶花园的长形（Long）酒吧和天空（SKY）酒吧的特色体现了大都市的氛围、360°全风景及新潮音乐。水疗中心的设计语言简洁、前卫，这里配备有水疗法的室内加热泳池、极可意浴缸、桑拿房、日光浴室、美发沙龙、装备齐全的理疗房和按摩室。

严谨的内部装饰、通过直边简洁剪裁的家具，结合艺术装饰品和动感的装饰画表现出来。每一处内饰都具有相同的设计手法，根据其功能把它的密集度最大化和最小化。在通往卧室的走廊中，透过结构化的木板形式感受到了体量和温馨。进入卧室，就能看到三种主要材质：木材、油漆和玻璃。这三种修饰材料与平滑的布艺相结合，制造了豪华又亲和的空间氛围。只需环顾大堂，你就能看到高端的饰品和家具。

MET酒店是当代设计的示例，它所具有的形态、氛围、豪华、性感和舒适感，使客人身临其境并惊叹该酒店的“独一无二的设计”。

THE MET HOTEL

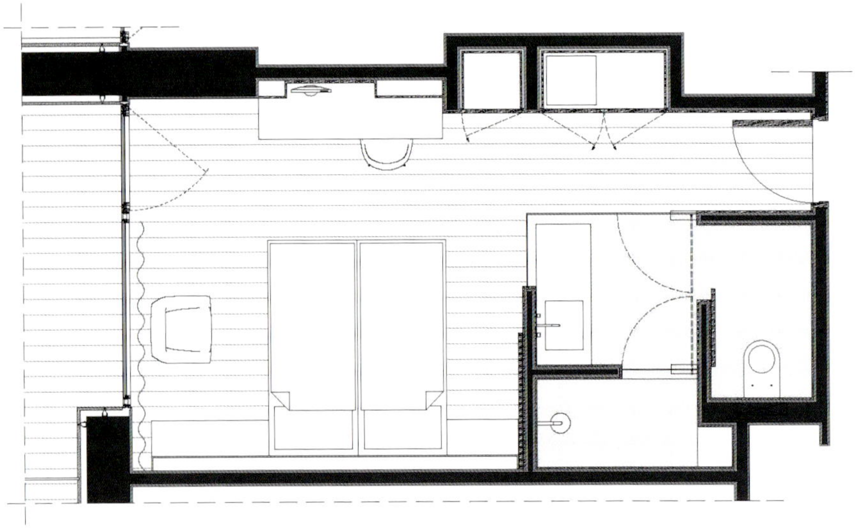

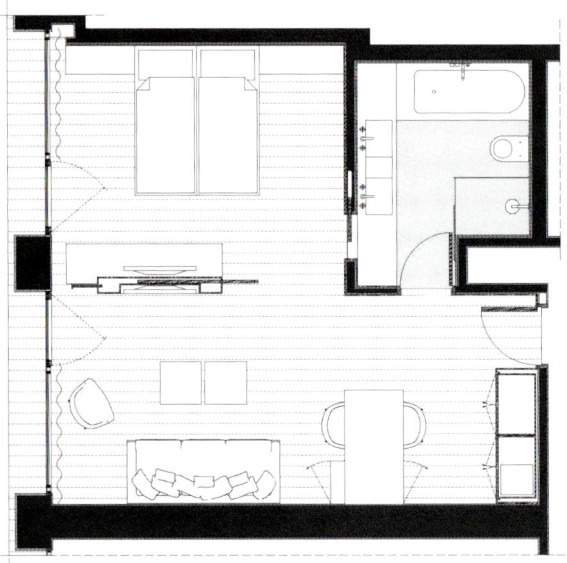

遐想

Reverie

土著民族酒店–博物馆
Hotel-Musée Premières Nations

设计公司：LEMAYMICHAUD建筑设计公司

项目地点：加拿大魁北克

项目面积：4850平方米

项目摄影：Pierre Bélanger

This project is a unique alliance between 4-star accommodations and a museum experience that takes visitors on an exceptional voyage through the traditions and contemporary culture of the First Nations and Inuit. Located in an enchanting setting on the Akiawenrahk River, the hotel/museum complex evokes the architecture of the longhouse, traditional dwelling of the Iroquois people, which include the Huron-Wendat Nation. Combining contemporary comfort and cultural experience, the architecture warmly welcomes guests to its 55 4-star rooms and suites, worthy of the legendary hospitality of the Huron-Wendat. The evocative location adds a special cachet to business meetings and receptions, along with the authentic flavours of gourmet cuisine inspired by the First Nations, a 125-seat restaurant with a private dining room and 5 meeting rooms that can accommodate up to 140 people.

The museum shares some common spaces and services with the hotel. Its cone shape is reminiscent of an Algonquin smokehouse and the hospitality of a hearth shared by all nations. The design also symbolically evokes a lowered stockade, a place where peace and harmony reign and where protective defences are no longer necessary. The two buildings therefore symbolize the two great cultural families of the East, the Iroquois and Algonquin. The museum presents a living history and artefacts imbued with a tangible memory. The interactive format displays a collection of rare objects central to the Huron-Wendat culture, exploring themes of territories, memories and knowledge.

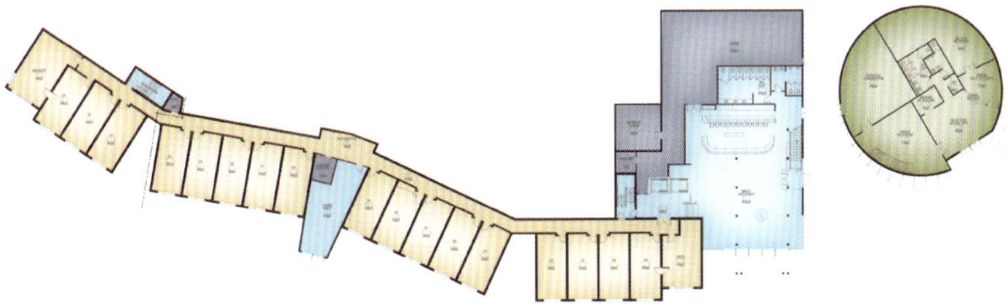

现代与传统的联姻
在圣查尔斯河畔
让人遐想

此项目是四星级酒店与博物馆体验的独特联姻，通过因纽特人的传统和当代文化的结合带给客人非同一般的旅程。这幢酒店和博物馆的混合体坐落于迷人的圣查尔斯河河畔，长形屋的建筑让人联想到易洛魁族的传统住宅，该民族包括休伦·温达特族。建筑将现代舒适感和文化体验融为一体，把客人们热情地迎进55间四星级客房和套房，名副其实地体现了休伦·温达特族的好客美名。建筑所处的周边环境引人遐想，连同原著民的地道风味美食一起为商务会议和接待增加了特别招牌。餐厅可容纳125人，包括一间包房、5间会议室，能容纳140人。

博物馆与酒店共用某些公共空间和服务。圆锥形的外观让人怀旧，阿尔冈昆族的熏肉室以及所有民族共有的炉膛般的热情好客带给人们更饱满的感受。设计还利用象征性的低围栏，寓意这是一个和平、和睦的地方，不再需要保护性的防护设施。因此，两个建筑物象征了加拿大东部的两大文化部族——易洛魁族和阿尔冈昆族。博物馆呈现了活生生的历史及浸润着有形回忆的人工制品。互动形式展现了休伦·温达特文化重要的珍贵收藏品，探索着地域、回忆和知识的相关主题。

豪华、宁静和新颖

Luxury, Serenity and Innovation

卡尔加里Germain酒店
Germain Hotel Calgary

设计公司：LEMAYMICHAUD建筑设计公司

项目地点：加拿大阿尔伯特省

项目摄影：Germain集团

宁静的感觉
闪烁在宜人的第九大道
呼唤来此的人们

Germain Calgary hotel : part of a new building of 22 storeys – 28,225 m², including 143 hotel rooms, 40 condominium units, 8800 sq ft of office space, a restaurant, 7 levels of underground parking and a room on the top geothermal basement. Le Germain Calgary offers travellers something new, compared to the other hotels in the group. "Every time we build a new Germain hotel, we strive to reinvent the Germain experience of style, comfort and personalized service while adding an innovative design element," explains Viateur Michaud, co-founder of the architectural firm.

The Germain Calgary boasts an exceptional location in the heart of downtown only a stone's throw from all the city has to offer. Connected to the city's network of elevated pedestrian walkways, the hotel, unlike other Germain hotels, actually opens directly on to 9th Avenue. The lobby features a two-story glass wall facing the main commercial street, thus creating a unique vibrancy between the street and the hotel space. The materials and interior design were selected to combine ease and comfort. "Our goal was to create a welcoming environment with a flash of innovation," explains Viateur Michaud. The hotel's exterior cladding of Prodema wood-laminate panels prefigures the warmth of the interiors, where wood is a dominant element. Beyond the impressive fenestration of the façade, the main entrance is a vibrant, active space where an immense glassed-in lounge open to the street takes up almost the entire lobby.

卡尔加里Germain酒店占据了一个22层新楼宇的一部分，这个综合楼的建筑面积为28 225平方米，包括143间酒店房、40个高档公寓单位，约820平方米的办公空间、一个饭店、7层地下停车场和地热地下室顶部的一个房间。卡尔加里Germain酒店与集团的其他酒店相比，为旅客们带来了一些新意。建筑事务所的创建人之一Viateur Michaud说道："每次修建一座新的Germain酒店时，我们都会努力重塑Germain酒店在风格、舒适度和个人服务方面的体验，同时加入新颖的设计元素。"

酒店地理位置优越，坐拥市中区的核心地段，矗立于第九大道和中心大街的街角，酒店距离城中的文化、休闲和商业区仅几步之遥与Germain集团的其他酒店不同，该酒店连接着城市的人行道网络，实际上直接朝向第九大道开放。大堂的特色是面向主要商业街的双层高玻璃墙，从而营造了大街与酒店空间之间独特的活力。酒店选用的材料和内部设计既休闲又舒适。Viateur Michaud说："我们的目的是创造一个闪烁着创新之举的宜人环境。"酒店外部覆层的Prodema木幕墙预示了酒店内部的温馨，木材是主要的内饰元素。除了外墙壮观的开窗，主入口更是一个动感活泼的空间，一个由玻璃包围、朝向大街的巨大休闲吧几乎占据了整个大堂。

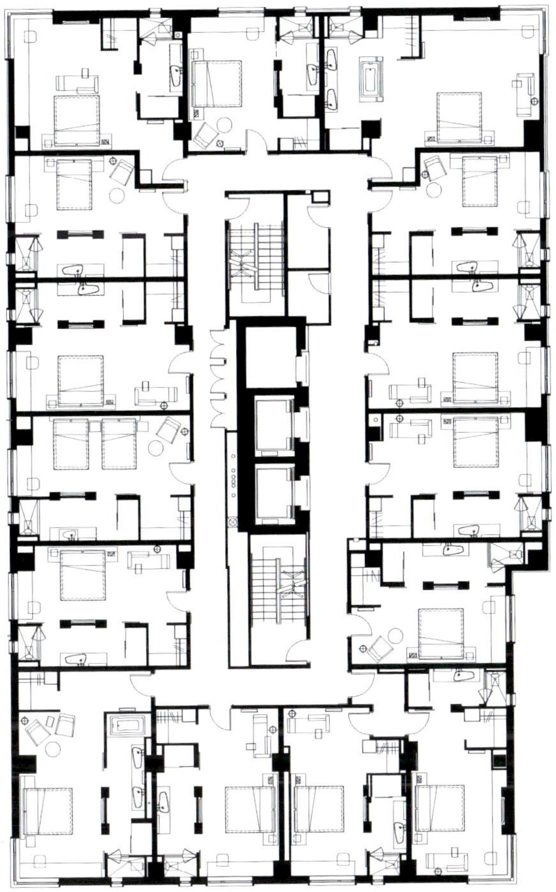

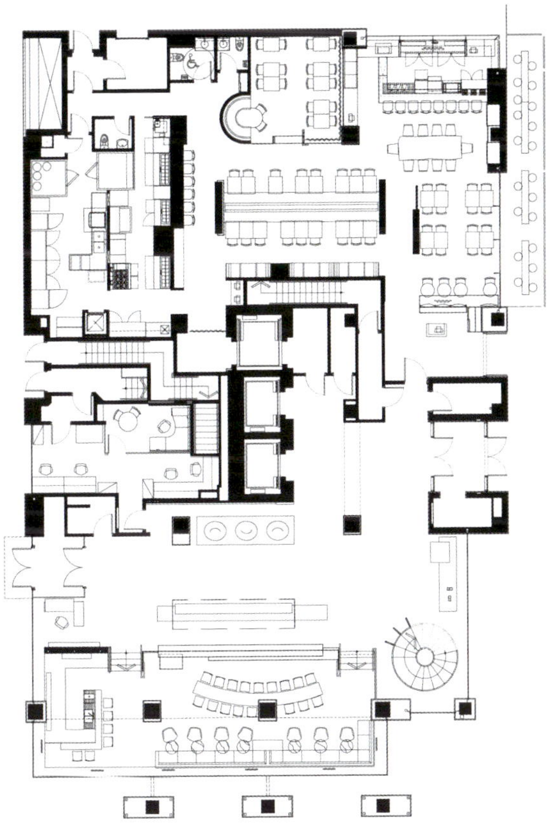

线条风情

The Flavor of Lines

蒙特利尔Germain酒店
Hôtel Le Germain de Montreal

设计公司：LEMAYMICHAUD建筑设计公司

项目地点：蒙特利尔

项目摄影：蒙特利尔Germain酒店

Beyond creating a pretty room, the notion of comfort has been the top priority in the design process, much like the first version of the Germain created 10 years ago. The lines and "flavour" of the Germain never cease to be present in this new creation, although turned towards modernity and future. Every room is designed in order to take advantage of the panoramic views of adjacent towers, the building stating a transition between the work space and the night universe.

The use of materials such as white lacquer, frosted mirror, nuances of chocolate, black and white are masters of this new environment. With the collaboration of Marie Saint-Pierre for the beddings, the rooms are enriched with a "haute couture" flare. Having created the reputation of the establishment, some key elements were judiciously preserved. Thus, the bathroom with its opening to the bedroom and the chocolate colour woodwork of origin, were restored and preserved. The goal was to achieve a new product which, without denying its origins, would add to the luxury, comfort and avant-garde image of the establishment.

New lounge chairs were created specifically for the rooms; covered with black wool jersey, they are located near a luminous table which also serves as ambient lighting. A 4 meter long wall unit integrates clothes closet, ironing and work areas as well as a mini-bar. All corridors and conference rooms were also revisited with the same white, black and chocolate thematic.

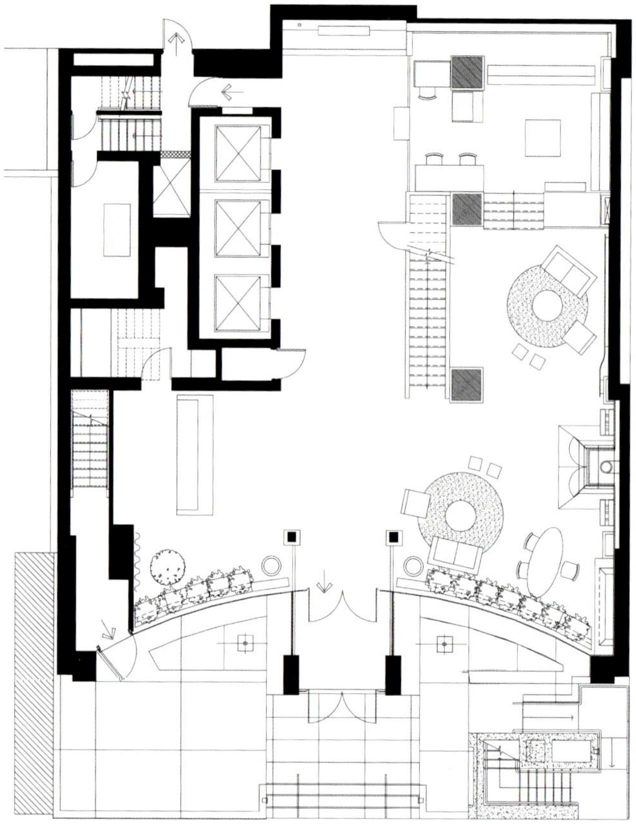

线条如同光芒
在建筑中延伸
在未来自由穿梭

除了建漂亮的房间，舒适概念在设计过程中一直占据着首要位置，这与Germain10年前修建的第一座酒店非常相似。Germain集团独有的线条和"风情"在创建这座新酒店时从未消失过，但是每个房间都设计成可以一览邻近的高楼大厦，这幢楼宇仿似工作空间与夜空之间的过渡。

白漆、磨砂镜、差别细微的种种巧克力色、黑色和白色主导着这个新的环境。配搭与Marie Saint-Pierre 合作设计的床上用品，房间里洋溢着"高级时装"般的气派。

鉴于该酒店已铸造的声誉，设计师明智地保留了某些关键元素。因此，开口朝向卧室的浴室和原有的巧克力色木制品得以恢复、保存。目的是在不否定原始风味的情况下，创造一种新产物，提升酒店豪华、舒适和前卫的形象。

设计师专门为客房制造了新型休闲椅。它们包覆着黑色羊毛针织布，靠近一个发光桌，这个桌子也充当氛围照明。4米长的墙体整合了衣橱、熨烫区和工作区以及一个迷你酒吧。所有走廊和会议室也用同样的白色、黑色和巧克力色主题重新修饰过。

非凡的酒店
A Divine Hotel

葡萄藤酒店
The Vine Hotel

设 计 师：Nini Andrade Silva
设计公司：Nini Andrade Silva Atelier Design & Interior Architects International
项目地点：葡萄牙马德拉群岛丰沙尔
项目面积：1575平方米
项目摄影：Alma Mollemans、Henrique Ssruca

是优质的葡萄酒
如同血液一般
流淌着灵魂

A hotel with soul, just like a good wine. The Vine is sophisticated and luxurious. With a touch of simplicity and discretion, just the way you like it. The Vine Hotel was created on the concept as the name implies. Arriving at the hotel, the great purple staircase and metal mesh symbolize the summer rain, offering a warm welcome.

In the lobby, there's a garden with sofas like big baskets of harvest and the purple cushions, the color of the wine. The gnarled branches of the lobby lounge and the mulled grape guest room draperies are meant to set travelers thinking about the region's rich viniculture and the exuberant wild vegetation. The Terra Lounge Restaurant represents the vineyards. All the floor finishing is covered with small round pebbles, representing the unique pebbled beaches of the island. The bathroom facilities are in form of a grape, as if each box was a grape. Four floors of rooms color-coded to reflect the four seasons – green, grey, purple and brown – are the dominating colors.

Each floor gives the sensation of changing season and follow close the grape's ripeness stage which later gave body to a great beverage.

葡萄藤酒店犹如一瓶醇香优质的葡萄酒，血液中流淌着它的灵魂。葡萄藤酒店集繁复与奢华于一身，将些许简单的精髓和些许周到的考虑都融入设计之中。投其所好，满足客人对视觉和感觉的双重要求。酒店的创意如同它名字的含义一样。到达酒店时，巨大的紫色台阶和金属网在欢迎你，宛如夏日瓢泼大雨般热切。

大厅中有一个花园，其间摆放着沙发，如同一个收获果实的大篮子，还有和葡萄酒一样颜色的紫色窗帘。大厅休息室里，粗糙树枝和葡萄酒红色的客房窗帘都可以让旅行者联想到该地区兴旺的葡萄酒酿造业和丰富的野生植物物种。Terra 休闲餐厅是葡萄园的象征。整个的楼层表面布满了小块的圆形鹅卵石，就像一片围绕小岛的独特鹅卵石海滩。浴室的设备都是葡萄形的，这里连每个盒子的形状都像一颗葡萄。4个楼层房间的色彩反映了四季——绿、灰、紫和褐，是各楼层的主色调。每层楼都传达着四季更替的感受。很快就到了葡萄成熟的季节，意味着你即将品尝到美妙的葡萄饮品。

剧院酒店
Theater Hotel

设 计 师：Nini Andrade Silva
设计公司：Nini Andrade Silva设计 & 内部建筑国际咨询工作室
项目地点：葡萄牙波尔图市
项目面积：5047平方米

As a true concert hall, the bronze tones and simply textures embody all the interiors. The main entrance offers us an eloquent poem from one of the greatest Portuguese Poets of the century - Almeida Garrett. As the main door opens we are pulled into the box-office to buy our ticket granting us the access into the rooms.

The restaurant is made from a palette of materials making its luxurious design possible, as a dream carved into reality. The furniture's design is a fusion of different design classics as well as a few surprises up their sleeves! The environment experienced in the bar takes us straight to any theater or opera, where the sofas design is inspired in the sixties and the countertops are made of recycled glass. Throughout all the space there are stage ropes as well as accessories and clothes belonging to famous artists who participated in the Filipe La Feria shows, a famous Portuguese theater director.

Penetrating the walls and on the way to the rooms, the corridors are coated by large curtains in order to give intimacy and coziness throughout the hotel. The 74 rooms - Gallery, Tribune, Audience, Junior Suite and Suite - spread over the six floors, are comfortable and sophisticated with shades of gold and bronze.

As a permanent search between the space harmony and design, Nini Andrade Silva Atelier has created several furniture pieces totally inspired into the theater and performing arts world, such as the sofa in the reception which represents a film reel of Footage Theater with images from Filipe La Feria shows, scenic lamps, chairs similar to those used in ancient theaters. All over the hotel it is possible to find accessories and theater clothes belonging to famous theater artists and stand out the beautiful work on recycled glass in the bathrooms facilities.

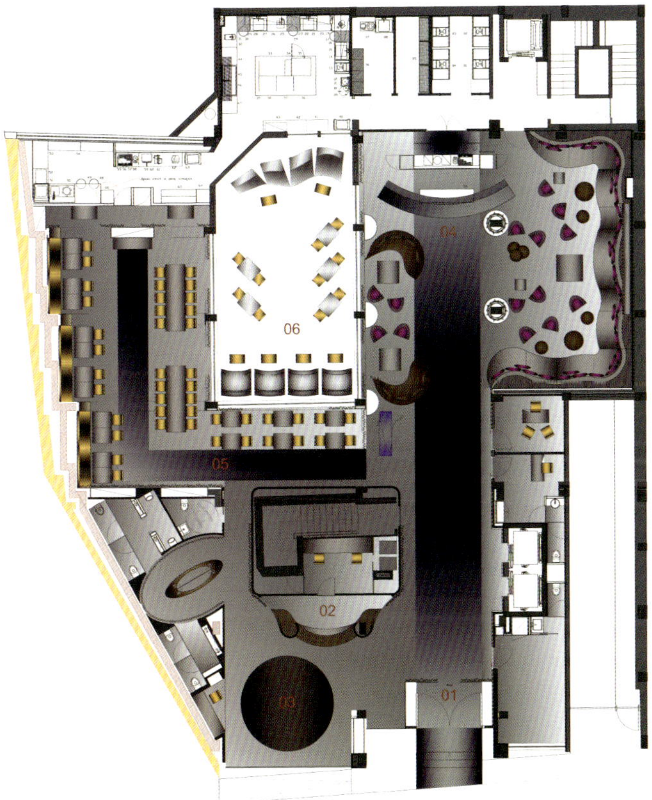

06
05
02
01

金碧辉煌的乐府
跳跃与空间的每寸肌肤
这是诗人的召唤

该酒店就像真正的音乐会大厅，青铜色和简洁纹理是全部内饰的主打特征。主大门上雕刻了葡萄牙伟大诗人Almeida Garrett的一首动人诗篇。主大门一打开，人们即被引到售票处购买进入客房的票据。

用多种材料构建的饭店尽显华丽气派，仿如梦境雕刻成了现实。家具的设计融合了不同经典元素及一些新颖造型。酒吧的环境体验顿时将人们带到剧场或歌剧院，这里的沙发设计灵感来自20世纪60年代，吧台用回收玻璃制成。整个空间随处可见舞台绳索以及著名表演艺术家们的配饰和衣服，这些艺术家都参演过葡萄牙著名戏剧导演Filipe La Feria的作品。

通向各房间的过道均装上了大遮帘，让整个酒店显得私密、安逸。分布在6个楼层的74间客房——走廊、看台、小型套房、豪华套房，舒适而优雅，闪烁着金色和青铜色的光泽。

为了达到空间和设计之间的永久和谐，Nini Andrade Silva工作室创作了几件完全受剧院和表演艺术世界启发的家具，比如接待处的沙发，象征了一卷连续镜头的胶片，上面是导演Fillpe La Feria作品里的一些画面，还有类似于旧戏院里的舞台台灯、椅子等。在整个酒店里，可看到属于若干著名戏剧艺术家的配饰和戏服，浴室里用回收玻璃制成的台面上摆放着这些精美的艺术品。

303

02
02
01
05

01
02
02
01
01
01
01
01

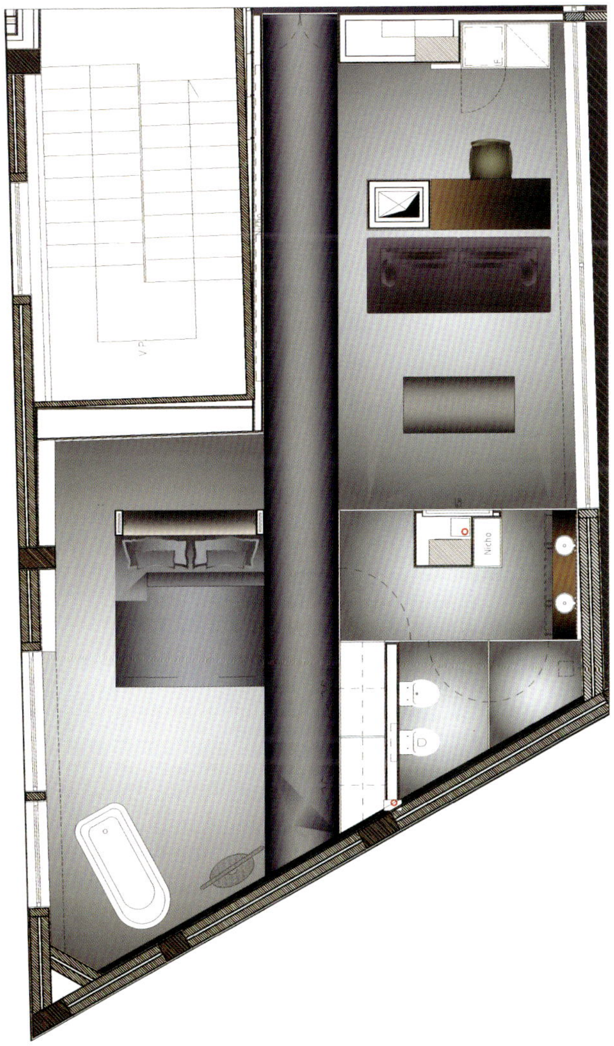

都市公园风情

Urban Park-style Hotel

喷泉公园酒店

Fontana Park Hotel

设 计 师：Nini Andrade Silva

设计公司：Nini Andrade Silva设计&内部建筑国际咨询事务所

项目地点：葡萄牙里斯本

项目摄影：Simon Frederick

Everything revolves around the name of the hotel – Fontana Park – an urban, cosmopolitan and sophisticated park. It is an urban hotel with contrasting, refined interiors. It has Ms Nini's signature with her 'Ninimalist' inspiration, contrasting design pieces with its enormous century-old roots transformed into installations. The social areas are very spacious, with a garden at the rear of the building.

The hall is the most monumental space, due to the height of the ceiling, yet it remains a welcoming and attractive space, with many interesting details. This is where everything starts. The reception area incorporates various iron elements in homage to the building's original function as a small factory for iron structures.

Windows generous in size and number assure that visitors are in constant communication with the dynamic urban landscape surrounding them, while an interior open-air courtyard – complete with waterfall and bamboo trees – references the green side of Lisbon. As the name itself suggests, nature and the park itself are the central themes, explaining why, aside from the use of black and white, the main color used is green. The criteria for the choice of colors were simple.

切都围绕着酒店的名字而展开——喷泉公园，一个极具大都市风情的精致公园，空间内部装饰精美、对比元素浓厚的都市酒店。该酒店颇有建筑师Nini的招牌风格——“Nini式极简主义“，把这些设计元素与文化历史共同融合到各种设计或装置中。社交区极其宽敞，酒店的后面有一个花园。

大厅因为天花顶的高度，成为最壮观的空间，但仍不失为一个欢迎宾客、吸引顾客的空间，其中有很多有趣的细节。一切正是从这里开始。接待区综合了诸多铁元素，以此向建筑的原有功能致敬：一家铁架结构的小工厂。

数量众多、宽敞大气的窗户确保顾客与周边的动感都市风情持续互动，而内部的露天庭院拥有瀑布和竹林，则代表了里斯本绿意盎然的一面。顾名思义，大自然和公园本身是酒店的中心主题，这解释了：除了黑白色之外，选用的最主要的颜色为什么是绿色。颜色的选择标准很简单。

一切从这里开始
都市的夜色
是公园的气息

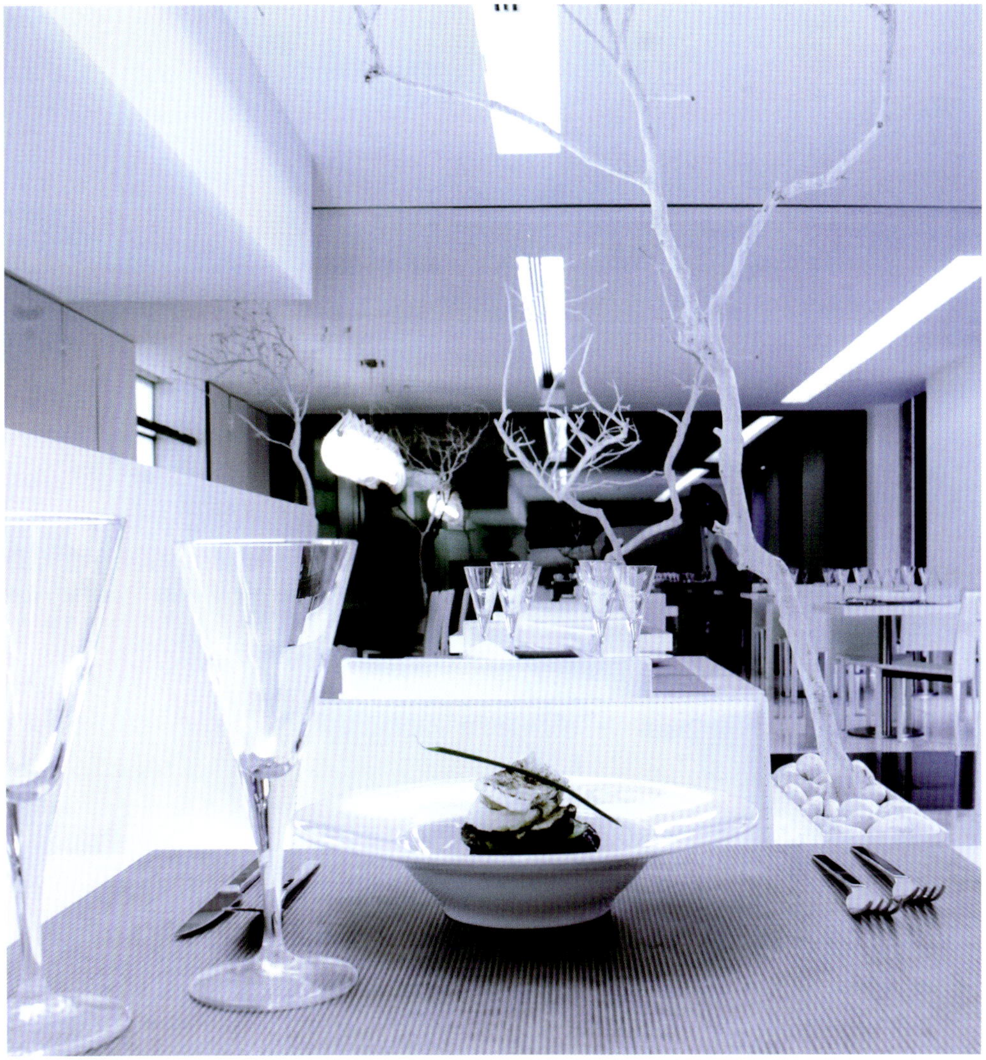

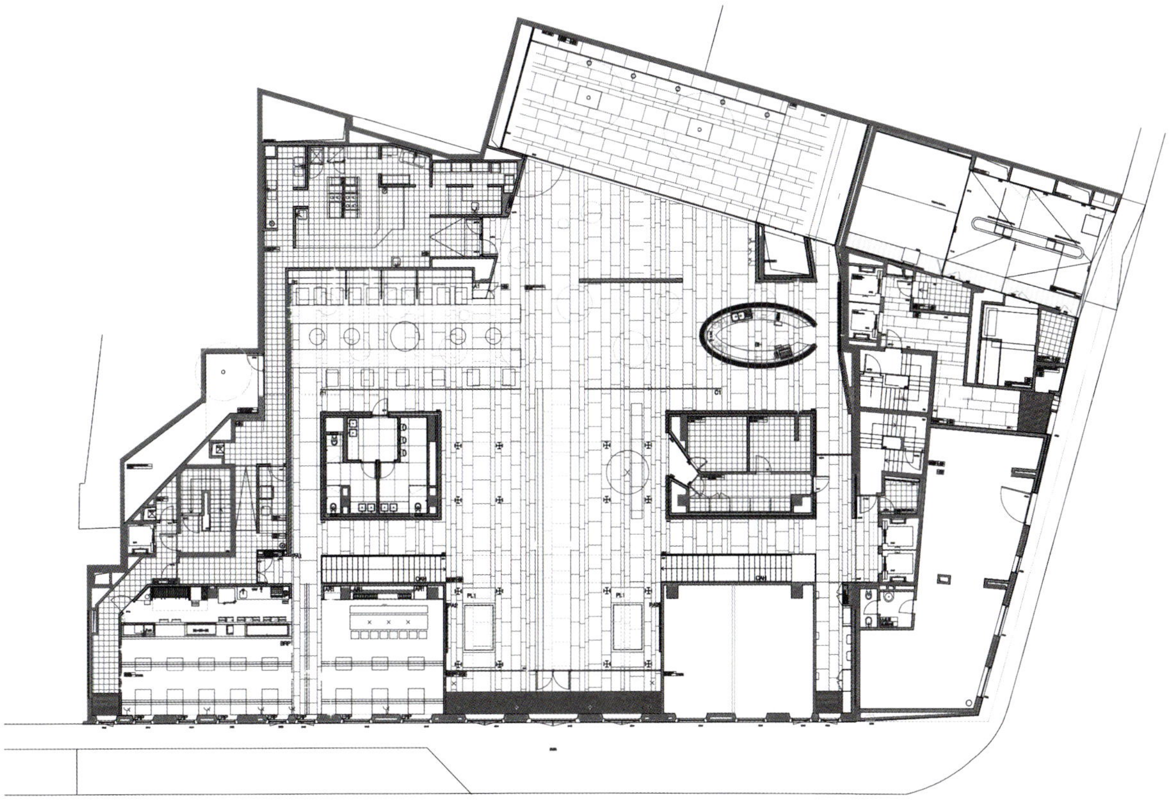

水艺

Art about Water

名都清水湾休闲商务酒店

Mingdu Clear Water Bay Leisure Business Hotel

设 计 师：孙志刚

设计公司：沈阳大展装饰设计顾问有限公司

项目地点：辽宁省沈阳市

项目面积：21 000平方米

主要材料：法帝米黄理石、月光米黄、热转韧条、琉璃

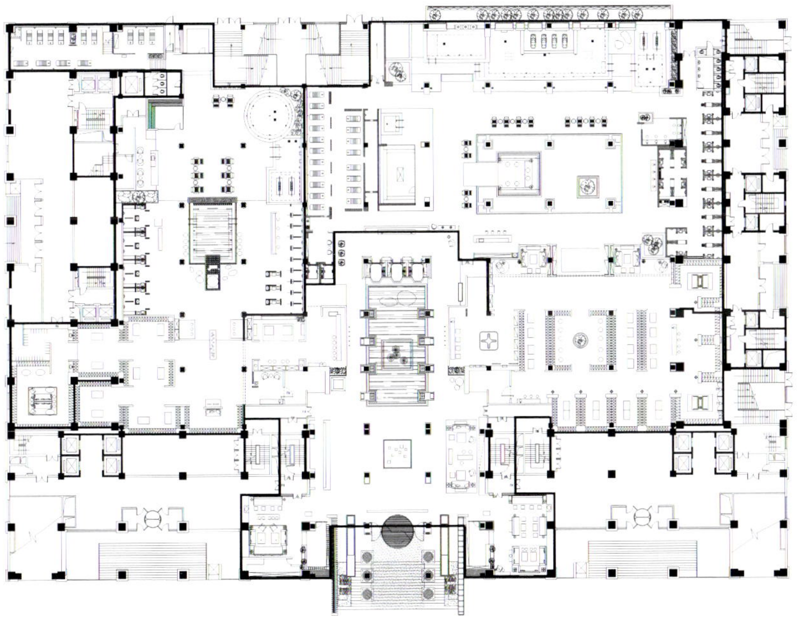

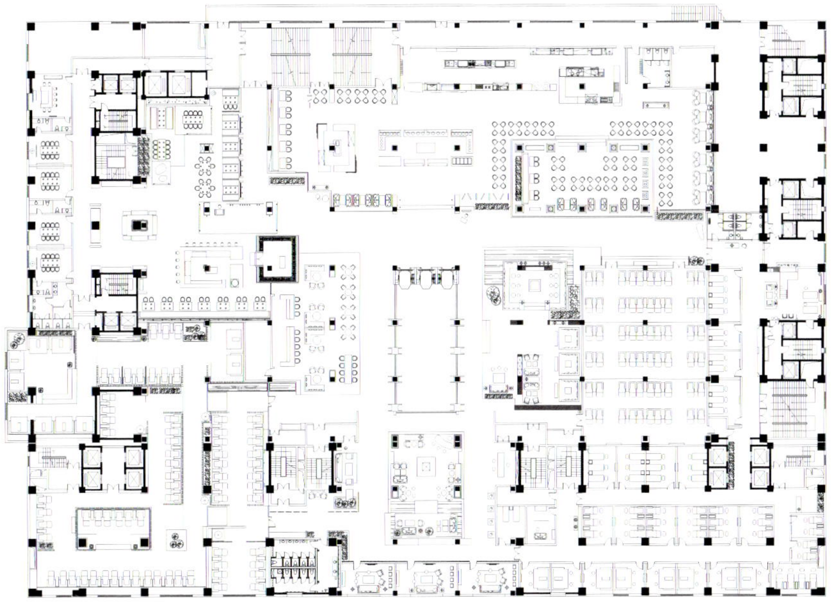

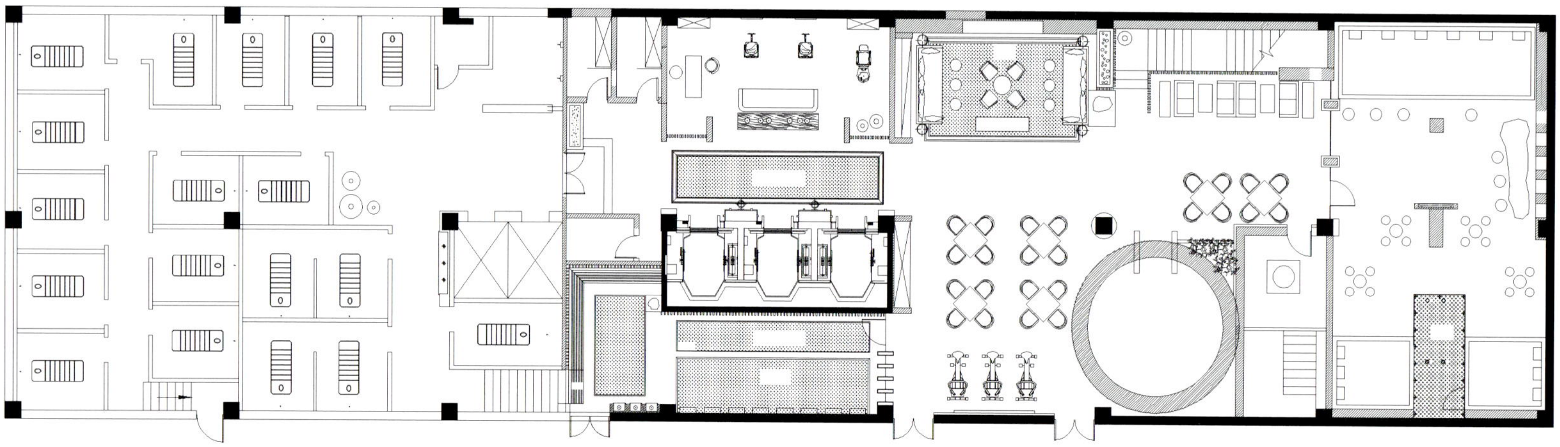

Designed with the theme of tranquil beauty, "Mingdu Clear Water Bay" Leisure Business Hotel brings a strong, heavy visual shock while its cultural taste is reflected from the elegance and restraint. The design is more focused on people's psychological feelings and emotions, trying more to reveal a sort of culture and spirit. Every link of the project, whatever decorative design or business concept, is guided by the industrial pattern of star hotels. The project not only upgrades the ideas of bath club available on market, but also makes people experience the new meanings offered from the club to the hotel, from the environment to the humanities.

"Water", the leading role in the design, drives the design and atmosphere of space from functions to views with the purpose of using colorless purity to contrast with colorful elements and integrating the elegance, refinement with the nature to blend every guest into the beautiful picture carefully planned by the design.

While "calmness", the obvious symbol running through every space, a strong force that is mature, discreet and knows tasting a life leads guests to a place like an artist's hall for them to appreciate the precious collections of the owner. Joyful spirit, cheerful visual sense and delightful sensations are design requirements for the hotel, also the people's increasingly enhanced pursuance once their basic needs are satisfied.

“名都清水湾”休闲商务酒店的设计以宁静作为美的主题，以浓郁和厚重形成对视觉的冲击，以典雅内敛作为文化品味的手段。设计师在设计中关注的是人们的心理感受与感动，想体现的是一种文化和精神。项目中每个环节，无论是装饰设计，还是经营理念都以星级酒店产业模式为主导。不仅提升现有市场洗浴会馆的理念模式，更使得人们重新体验到从会馆到酒店，从环境到人文带来的新内涵。

在设计中，“水”占据着主角，从功能到景致，它带动着空间的设计和空间的气氛，用无色的纯净映衬多彩的元素，把一股高雅和精致同自然相结合，让每位来客都融入到设计所精心规划的美丽构图中。

“稳”是贯串各个空间的明显符号，一股成熟稳重的浓厚力量使人们懂得品味生活的真谛，仿佛把来客带到艺术家的厅堂，欣赏着主人的珍藏。精神愉悦、视觉愉悦、感观愉悦是酒店的设计宗旨，也是人们满足基本需求日趋提升的追求。

曲水流转
清雅自然相成趣
名都艺天堂

奢华品味 尊崇感受

Respect the enjoyment

大连国尊王朝大饭店

The Dalian Copthorne Hotel Dynasty

设 计 师：殷艳明

设计公司：深圳市创域设计有限公司

项目地点：中国大连

项目面积：20 000平方米

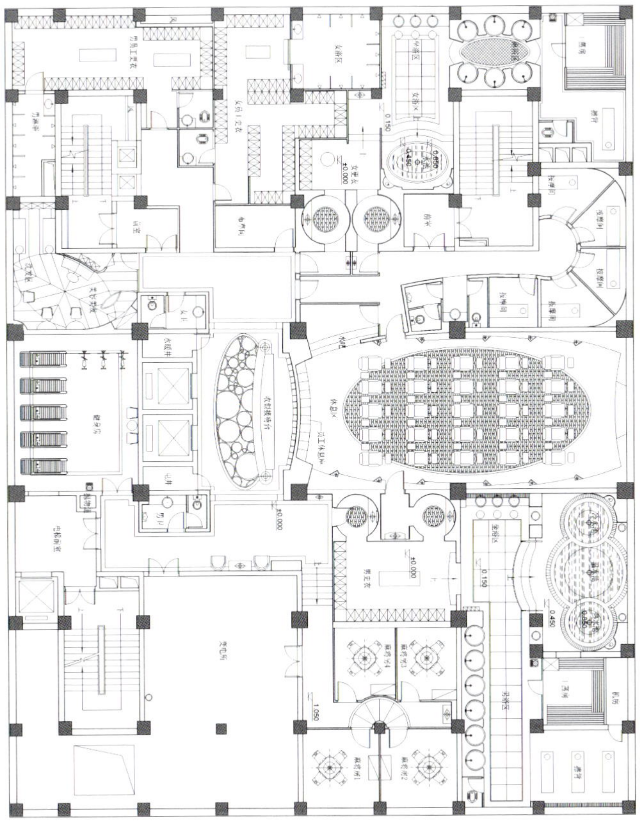

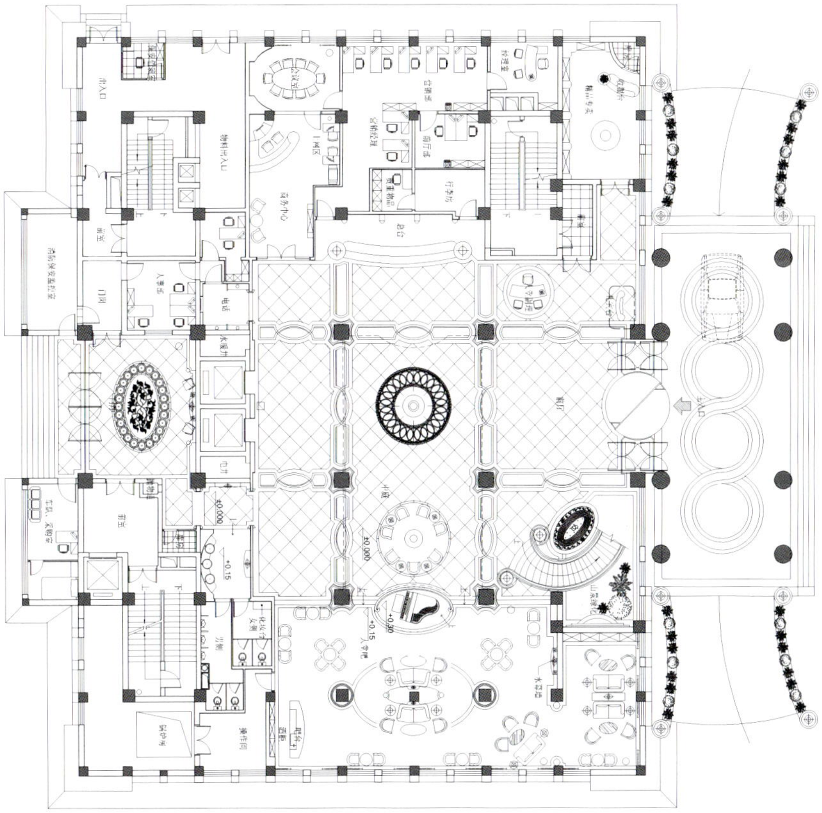

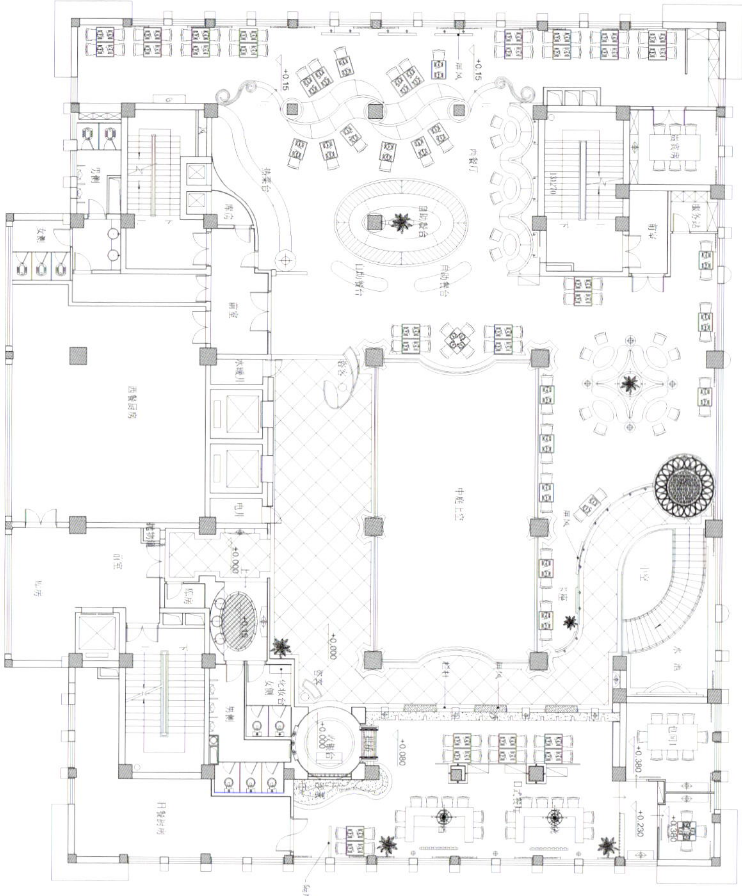

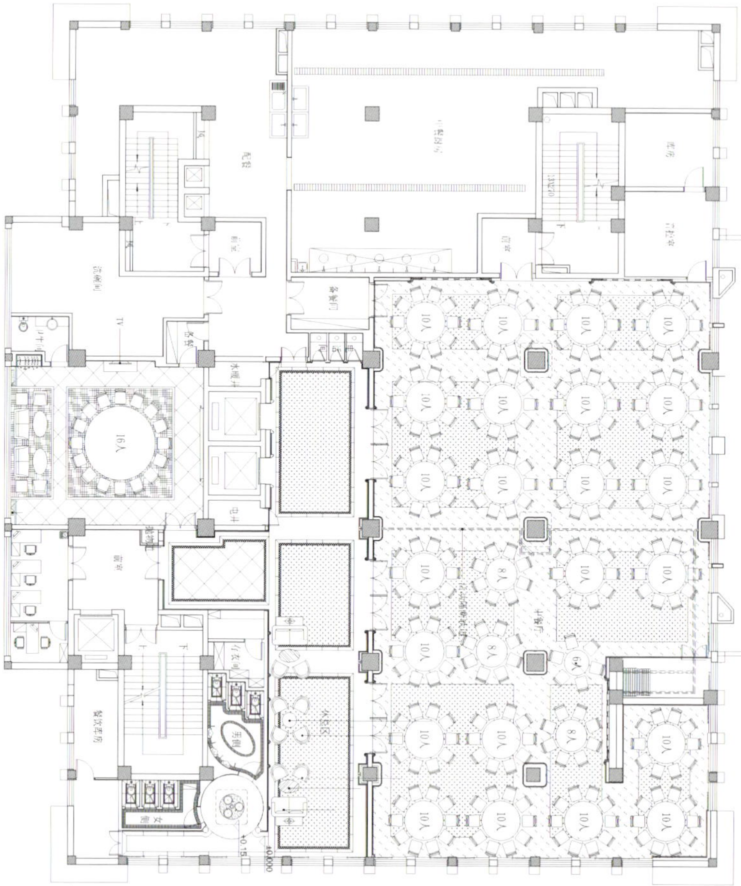

海洋生态的背景下
是新古典主义风格
也是空间的海洋情怀

In the design features, the grasp of the most pure of space design concepts, focusing on different areas of personalization, using the latest material to express a blend of traditional and modern ideas. At the same time, the concept of interior design continues the unified design style, in the integration of the perfect combination of form and function at the same time, pay attention to color matching, color outlining the overall harmony, the rich luxurious feel of space and generate a strong visual effects. Compact flow line and comfortable spatial scales is given to only use club-style, different from the traditional star hotel room, through the clever combination of art and jewelry, in order to make room for the spirit of humanism is more humane.

As a premier five-star boutique hotel in Dalian, Dalian Copthorne Hotel Dynasty furniture, accessories, materials in each luxurious room are carefully selected Dalian Copthorne Hotel Dynasty make the greatest possible luxury to provide a comfortable space for guests to use. Hotel executive floors for senior business offers a range of exclusive facilities and supreme luxury of personalized service. The hotel's dining facilities ranging, sophisticated Western-style restaurant serves Western cuisine, elegant Japanese-style restaurant offers authentic Japanese cuisine, rich oriental flavor, rich supply of delicious Chinese food lunch private rooms, let the guests to fully experience the classical Chinese food culture. Lobby Bar offers a selection of drinks and snacks, and accompanied by live bands playing light music, and unforgettable is the House Floor Grill for the supply of all kinds of Western-style steak, barbecue and so on. With delicious dishes and a glass of red wine sitting on the highest point of the hotel to enjoy the city's beautiful scenery. The hotel also boasts a comprehensive business center, recreation center, sauna and entertainment center, a comprehensive fashion boutique brands so as to add a variety of leisure options for guests.

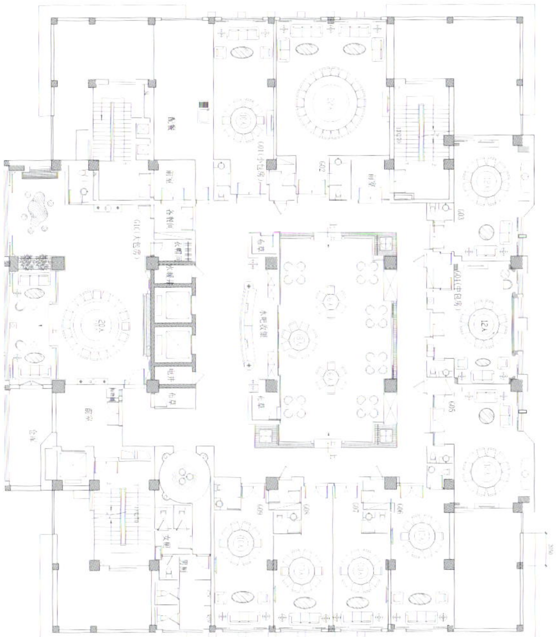

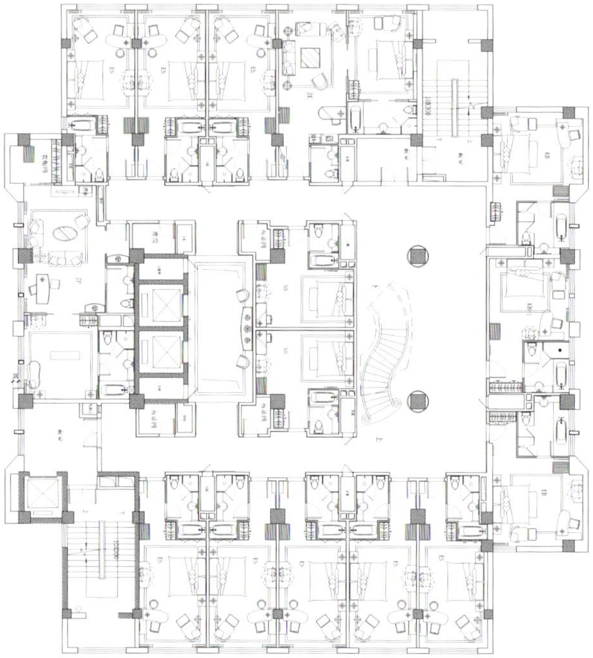

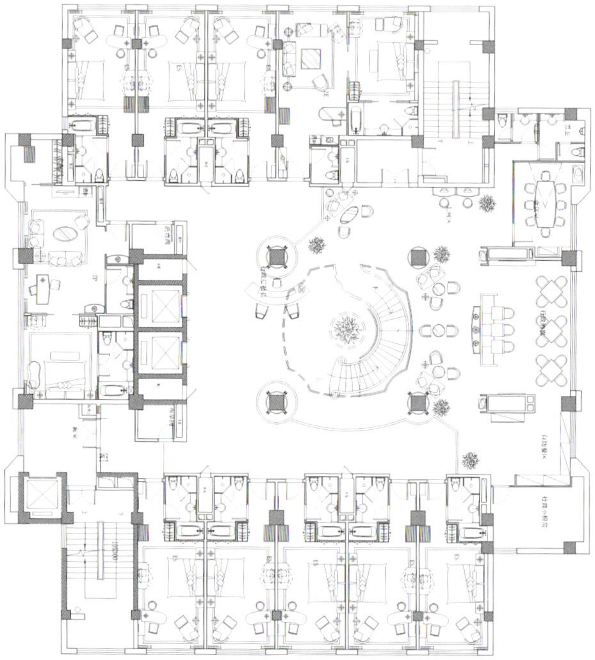

在设计特点上，运用最纯正的空间设计概念，注重不同区域的个性化，采用最新的素材去表达传统与现代交融的理念。内部设计概念延续了统一的设计风格，在造型与功能完美融合的同时，注重色彩的搭配，整体色调勾勒出和谐、奢华的丰富空间感受，并产生强烈的视觉效果。紧凑的流线和舒适的空间尺度给人独尊空间的感受，区别于传统的星级酒店空间。通过艺术品与饰品的巧妙结合，赋予空间以人文的精神内涵，让人流连忘返。

作为大连首屈一指的五星级精品酒店，每间豪华客房中的材料、家具和配饰都经过精心的选择，尽最大的可能提供一个舒适豪华的空间供客人使用。酒店行政楼层为高级商务客人提供了一系列独享的豪华设施和至尊的个性化服务。酒店的餐饮设施一应俱全，精致的西式餐厅供应西式美食；典雅的日式餐厅供应纯正的日式料理；富有浓郁东方色彩的中餐包房供应美味的中餐，让客人充分感受中国饮食文化的古典和精致。大堂里提供各种精选的饮料和点心，伴着现场乐队演奏的轻音乐，更让人流连忘返的是阁楼层扒房所供应的各式西式牛扒、烧烤等。品着美味佳肴，端一杯红酒，坐在酒店最高处欣赏着城市的美好景色。酒店还设有完备的商务中心、康体中心、桑拿娱乐中心、综合性时尚名牌精品橱窗等，为居住在此的客人增添多样的休闲选择。

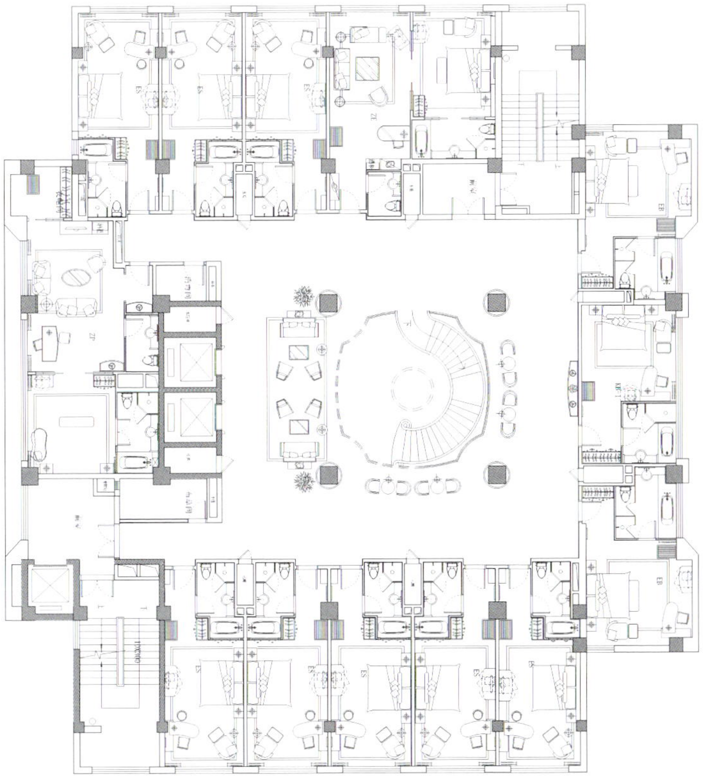

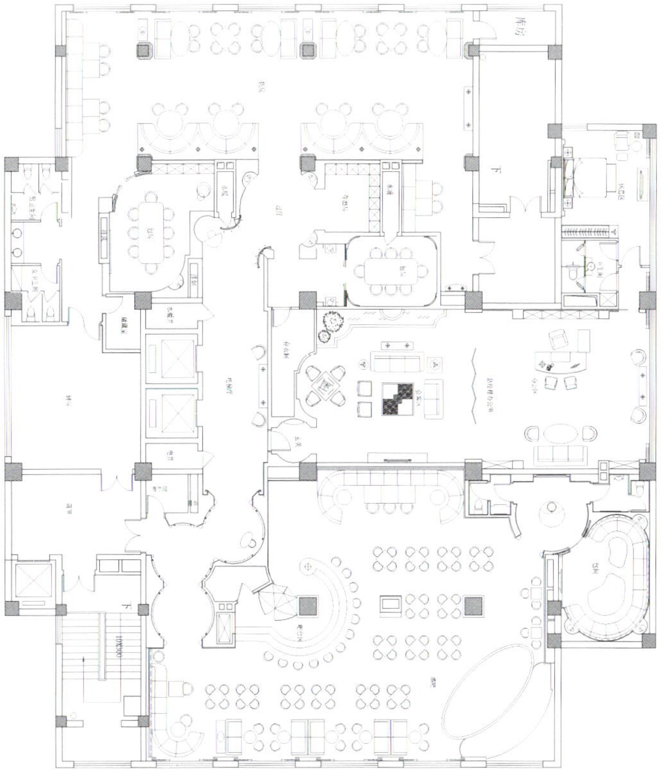

幻境

Fairyland

观止二二
Villa Motel

设 计 师：郑唐皇

设计公司：郑唐皇计划设计室

项目地点：中国宁波

Villa Motel features a number of unique, fashionable suites whose interiors are designed with respective themes and natural landscaping like nature itself, they look so beautiful and gorgeous that people feel like being in a fantastic beauty. The space is clean in lines, but good at creating a stylish romantic feel, where sumptuous, elegant furnishings and equipment show class, offering any consumer a romantic, warm, safe and intimate space while the bath being an enjoyment in which people can do as he pleases, basically each room is equipped with a large massage tub, plus the steam evaporated from RO pure water, making your body, heart and soul bathed in the emancipation of snatching a little leisure from a busy life.

The architectural exterior breaks through the seriousness of a usual square structure, making the establishment look like a sculpture in the city through the origami-like shape, whose clean, crisp lines stand out in the city's incessant stream of cars and pedestrians. Its dark clapping presents a low-key steady air without losing a graceful charm.

Unlike the conventional enclosing walls or fences to strongly define the scope, the surrounding landscape is shown using digital grass slopes. The irregular folded-board like facades enliven the space and the lawns are approachable, also echoing with the building dimensions, achieving a uniform effect.

如同置身幻境的感觉
焕然天成的自然造景
让人徜徉温馨之中

观止二二酒店内有多款独特的时尚套房，房内均采用不同主题的设计，还有浑然天成的自然造景，美轮美奂的感觉仿如置身于幻境般的美感当中。其空间线条简洁，却擅于营造时尚风情的浪漫情调，华丽高雅的装潢与设备呈现高贵的气质，让顾客有一个浪漫、温馨、安全且隐秘的私人空间，而沐浴是一种随心所欲的享受，每套房间基本配备大型按摩浴缸，另有由RO纯水散发而出的蒸气浴，让您的身、心、灵享受“偷得浮生半日闲”的解放。

建筑外观打破方正格局给人的严肃感，利用折板包覆量体的手法，使建筑物仿佛成为城市中的雕塑品，简洁利落的线条，在都市车水马龙的景象里独树一帜，而深色的外观材料，则表现出低调稳重的气质，却不失高雅的韵味。

景观部分以数字草坡的手法来呈现，有别于传统使用围墙或格栅强势界定范围的手段，不规则的折板使得空间变得活泼，而草皮也使人感到亲近。此外，景观与建筑量体形成呼应，让整体效果更一致。

CCTV

清新城市空气

Fresh Air of City

清新酒店
Fresh Hotel

设计公司：zege设计公司

项目面积：6000 平方米

主要材料：塞茵那石、三色玻璃、可丽耐、三色玻璃、石膏板

Maximized by sacrificing only 3 rooms from the old building, 3 suites and 6 executive rooms with private Zen gardens were created as well as facilities and areas that did not exist and unique aesthetics to all the communal areas was given. It is notice that the transformation of the old spaces to the unique and excellent new ones in the pictures. A colorless, uninteresting old building was turned into a cool and sleekly designed hotel. Areas such as a business center, fully equipped gym, sauna, steam bath and conference rooms were created in order to enhance the customers visit in the best possible way.

Moreover, the roof top which used to be just a swimming pool area operating in the morning hours was redesigned into a lounge bar, an a la carte restaurant with a new kitchen and the swimming pool area.

Furthermore, the hotel was built in the heart of Athens that old classical buildings and graphic streets surround it and where ordinary city hotels are found in every corner of the city. Fresh Hotel came to bring a fresh breath in city hotels with an unconventional atmosphere and an architectural example. While trying to create a memorable hospitality experience, it enhanced tourism and upgraded the area, at the same time a sanctuary was created in a hotel that takes your breath away.

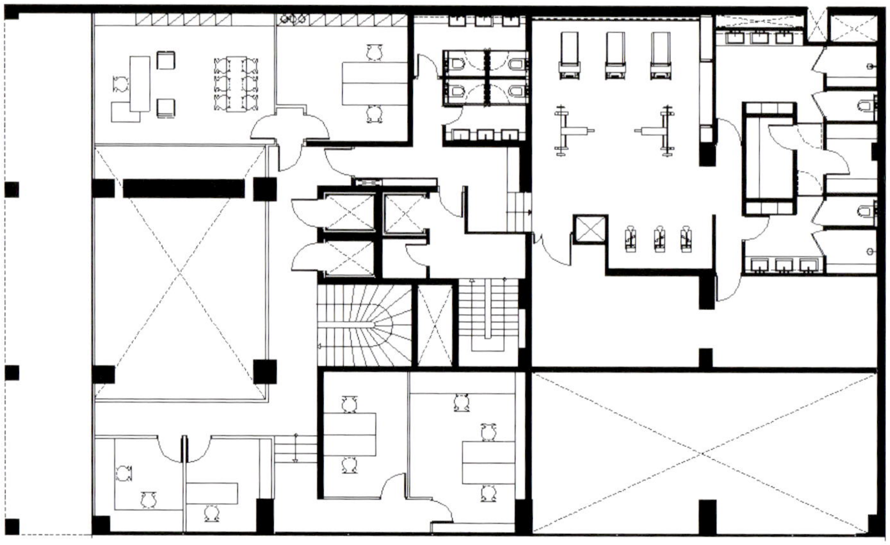

缤纷的色彩
亭亭玉立的两支向阳花
当阳光洒进来
世界是清新的

酒店的每个角落都得以利用，具有了一定的功能，创造了空间的实用价值。旧楼里的3个房间最大化了房内空间，建造了3个套房、6个有私人禅式花园的行政房，还增添了之前没有的设施和区域，并且赋予全部公用区域地方独特的美感。从照片里独特的旧空间到出众的新空间，其转换效果实在令人惊讶。一幢无色、乏味的旧楼被改造成一个有型有格、设计时尚的酒店。在商务中心里，设施齐全的健身房、桑拿房、蒸气浴室、会议室等区域，尽可能以最好的方式提高客人的入住率。

另外，屋顶——过去只是一个在早上开放的游泳池，现在被重新设计成一个休闲吧、一个配有新厨房的点菜餐馆和泳池区。

还有，酒店位于雅典中心，周边是古老的经典建筑和风景如画的街道。在这里，城市的每个角落都能找到普通的城市酒店。清新酒店以其不同寻常、非传统的氛围和建筑样式给那些一般的城市酒店带来清新空气。一方面试图营造令人回味以及热情周到的体验，另一方面它促进了旅游业的发展，提升了地区形象，同时酒店里的圣殿让人叹为观止。

时尚族的经典情怀

Classical Feelings of Fashinable People

Mumuri 酒店

Murmuri Hotel

设 计 师：Kelly Hoppen

设计公司：Kelly Hoppen Interiors

项目地点：西班牙巴塞罗纳

Located in the emblematic Rambla de Catalunya, close to Las Ramblas, La Pedrera and surrounded by the best national and international boutiques, Hotel Murmuri offers those visiting Barcelona an incomparable accommodation choice.

The stone façade and the splendid windows of the building are in harmony with the typical architectonic style of the Eixample quarter. However, the interior has been designed in a wonderfully sober, pleasant and sensual fashion by the acclaimed British interior designer Kelly Hoppen who has managed to take care of every detail and to create a hotel with those personal touches that please even the most demanding and refined guests.

All 53 rooms have been decorated with an exquisite taste, meriting special mention the Suite with views over Rambla Catalunya and the Junior Suite. The combination of dark wood with various hues of beige, cream, grey and brown create a modern, contemporary and sophisticated atmosphere that will "age well".

Treading a thin line between modernity and tradition, Kelly Hoppen has made an intelligent use of neutral colours and rich textures managing to create a style that, though close to minimalism, it is luxurious enough to enchant even the most conservative guests. Her interior design projects have crossed international boundaries; she designed the Beckham's L.A house and even the first class cabins of British Airways.

The bar, Marfil by Murmuri, brings back a 1920's classic flavor that will be open all day and till 2 o'clock in the morning every day with a refreshing and light range of choices to be enjoyed both inside and on the privileged terrace. This will be the new meeting point for Barcelona's most fashionable people.

加泰罗尼亚大道风景
一道诱惑
是时尚的经典情怀

坐落于标志性的加泰罗尼亚大道，靠近兰布拉大道、米拉公寓，四周是顶尖的西班牙和国际性的精品时装店，Murmuri酒店给参观巴塞罗那的人们提供了无与伦比的住宿选择。

石头外墙、壮观的窗户与Eixample区的典型建筑风格和谐一致。但是，里面则由英国著名的内部设计师kelly Hoppen设计。处处显现得沉稳、愉悦并且极富感官刺激，设计中尽力处理每一个细节，创造出一个人性化，甚至能讨好最挑剔、最尊崇的客人的酒店。

值得特别注意的是，全部53间客房均装饰得精致有品位。豪华套房能看到加泰罗尼亚大道风景，小套房俯视欧陆式林荫道。深色木材结合不同色调的米色、乳白色、灰色和棕色，营造了经得起时间考验的现代、时尚、高雅氛围。

仿如用一根细线交织着现代和传统，Kelly Hoppen独具匠心地运用了中性色彩和丰富的纹理，营造的风格虽近似简约主义，但其豪华程度足以诱惑最保守的客人。其若干内部设计项目穿越了国际界限——曾设计过贝克汉姆在洛杉矶的家，甚至设计过英国航空公司的头等舱。

酒店的Marfil酒吧将带你重温20世纪20年代的经典情怀，酒吧全天开放直至次日凌晨2点，提供众多清爽、温和饮品，可在室内及专用露台享用。这将是巴塞罗那最时尚一族的新聚会点。

亲密接触
The Intimate Contact

格拉斯哥CitizenM酒店
CitizenM Hotel Glasgow

设计公司：Concrete Architectural Associates
项目地点：英国苏格兰

CitizenM Glasgow will be the third hotel to open in the citizenM family, and offers mobile citizens of the world affordable luxury in the heart of the city. The concept of the hotel is to cut out all hidden costs and remove all unnecessary items, in order to provide its guests a luxury feel for a budget price. Robust granite natural stone form a 8-storey building. The height, mass and firm appearance of the building are in close harmony with the context. A red colour glazed, double height facade marks the entrance.

CitizenM wants to contribute also to it's cultural environment. Two huge art works, printed on pvc mesh fabric are placed on the façade. In Glasgow, the artist Alex Dordoy expresses the open way of living of the mobile citizen in combination to the restless spirit of the city of Glasgow.

The ground floor houses commercial space with the entrance of the hotel on Renfrew street. Behind a red coloured glazed façade, a double height space links the Street with the world of citizenM on the first floor. A huge red coloured helical stair and a sky of hanging lamps welcomes you. The public area of the first floor is divided in several living rooms. Our purpose is to create a home-environment, by designing working areas, dining areas and sitting areas in cooperation with Vitra.

This bookcase contains, besides racks with magazines, newspapers and merchandise, also refrigerated glass cases with drinks, sandwiches, salads, sushi, typical breakfast and other kind of snacks and meals as well as fresh bakery products. At night, canteenM changes into a bar-area, parts of the bookcase are closed and different lighting scenes create an intimate atmosphere.

满天的吊灯
如璀璨的星光
螺旋状的楼梯
可以听见大海的声音

格拉斯哥的CityzenM酒店是CitizenM家族开设的第三家酒店，为移动的世界公民提供可以负担的、位于城市中心地段的豪华入住体验。酒店的概念是削减所有隐藏费用，取消全部不必要的物品，目的是以实惠的价格带给客人一种豪华感。刚硬的花岗岩天然石材组成这幢8层高的建筑，其高度、体量和强硬的外表与周围的环境非常和谐。一个双层高、镶嵌红色玻璃的外墙标志着酒店的入口。

CitizenM酒店也想为其文化环境出一份力。印在PVC网布上的两个巨大的艺术作品被放置在酒店外墙上。在格拉斯哥这座城市，艺术家Alex Dordoy结合该市的活力氛围，表达了移动公民开放的生活方式。

第一层是商业空间，包括位于兰弗街的酒店入口。红色玻璃墙后面的双层空间把兰弗街与第二层的CitizenM世界连接起来。巨大的红色螺旋状楼梯和满天的吊灯欢迎着你。第二层的公共空间划分成几个起居室。设计团队与Vitra家具公司合作设计了工作区、餐饮区和休息区，目的是制造家一般的环境。

这个20米长的书柜除了放杂志、报纸和商品，还容纳冷冻玻璃柜，该玻璃柜用于装盛饮料、三明治、沙拉、寿司等一般早餐和其他种类的小吃，还有膳食和新鲜的焙烤食品。晚上，canteenM自助餐厅则改变成酒吧区，书柜的一部分关闭，加上不同的照明情景，营造了亲密的氛围。

citizenM says:
drink me
TOM FORD

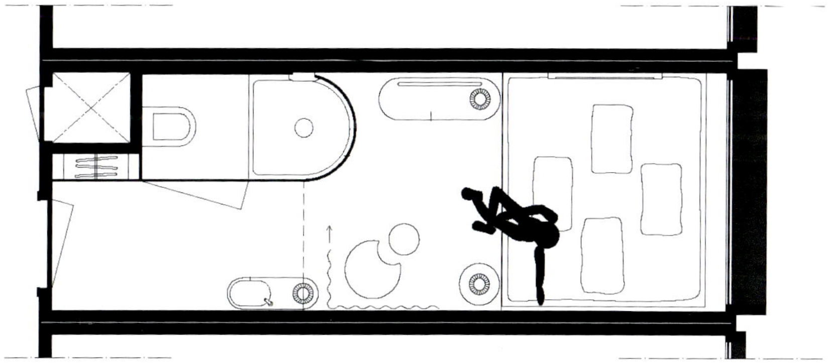

The World is a book, and those who do not travel read only a page.

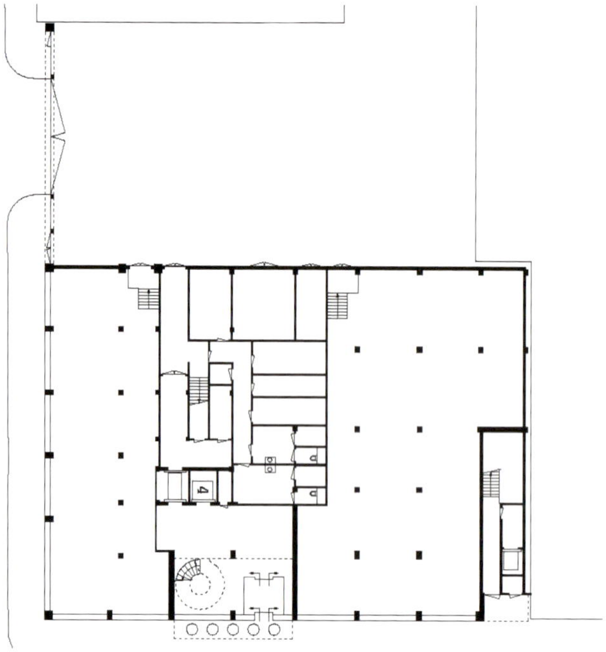
4

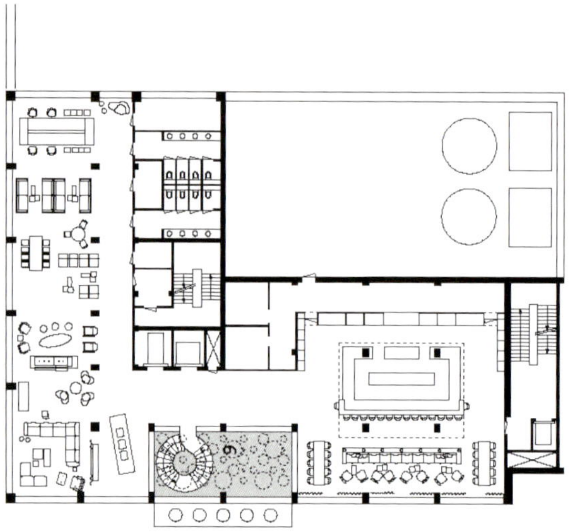
9

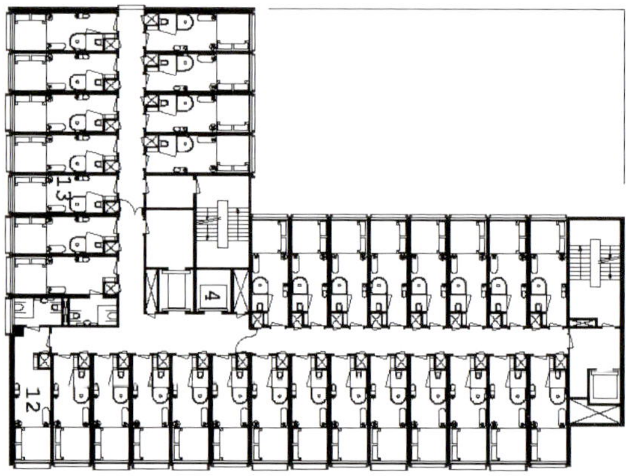
13
4
12

最经济的奢华

Economic Luxury

阿姆斯特丹市中心CitizenM酒店
CitizenM Hotel Amsterdam City Centre

设 计 师：Rob Wagemans, Erikjan Vermeulen, Jeroen Vester, Sander Vredeveld, Matthijs Hombergen

设计公司：Concrete Architectural Associates

项目地点：荷兰阿姆斯特丹

项目面积：6000平方米

citizen

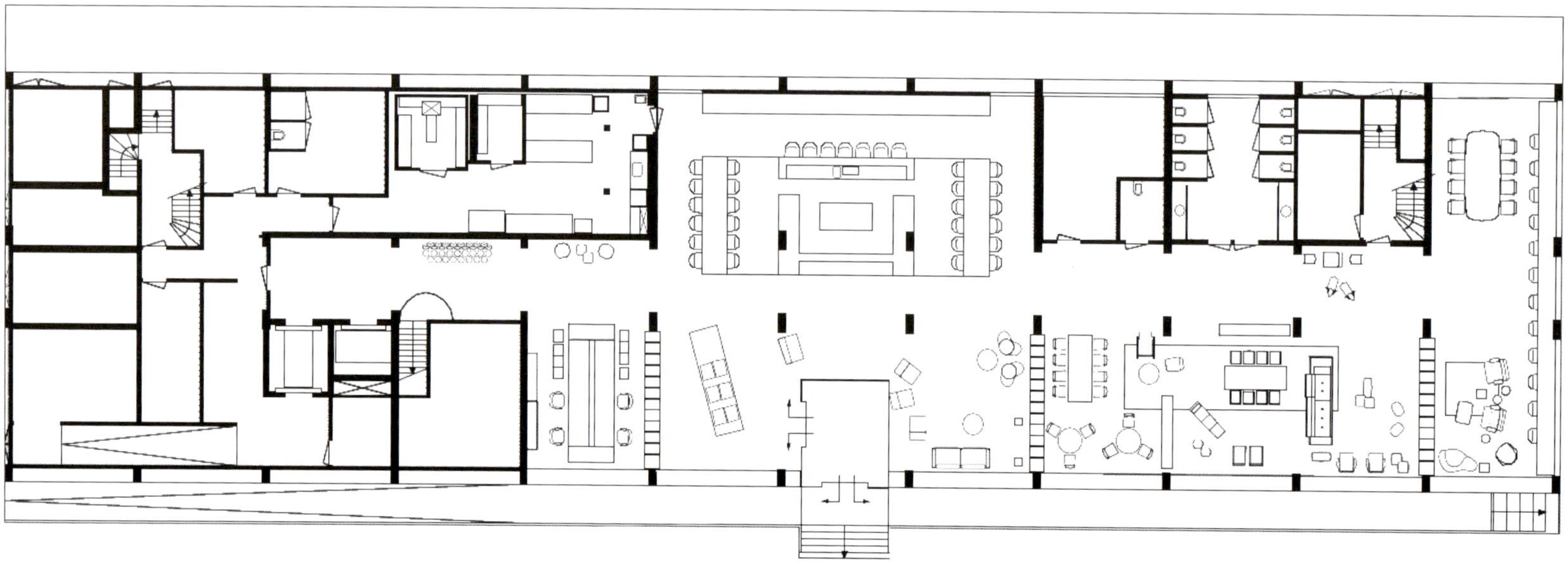

CitizenM is a new Dutch hotel group with concept of the hotel is to cut out all hidden costs and remove all unnecessary items, in order to provide its guests a luxury feel for a budget price. The hotel exists of 215 rooms of 14 m^2, all prefabricated produced in CitizenM's own factory and easy to transport. The hotel offers mobile citizens of the world affordable luxury in the heart of Holland's capital city.

A red glass box with electronic sliding doors allows entrance to the hotel. The hotel does not have a service desk but a table with six self-check-in terminals. One of citizenM's hosts is present to guide you through the checkin-procedure. The public area of the ground floor is divided in several areas that we like to call living rooms. Our purpose is to create a home-environment by designing working areas, dining areas and sitting areas.

The bed stands in front of a floor to ceiling, wall to wall window and is super-king size, 2.2 m x 2.0 m. It's white bed-linen and pillows offer a lounge area to watch TV on the flat LCD screen, together with the stuffed animal named Marvin (designed by Gewoon) that's placed in every room. Night tables on both sides of the bed contain plisse-shaded lamps. Underneath the bed is a huge drawer to store an opened suitcase or other personal belongings.

新成立的CitizenM 荷兰酒店集团本着减少开支的原则，省掉了所有不必要的家具，为顾客提供物美价廉的酒店服务。整个酒店由215个14平方米的小房间组成，全部由CitizenM自己的工厂里制造，方便了运输。在荷兰首都市中心，该酒店为“世界公民们”提供了最经济的奢华。

红色玻璃房子前是一面电控滑门。大堂未设服务台，只有6个入住登记处。一位大堂经理正恭候您的到来。底层的公共区域被划分为几个可以把它们称作“客厅”的区域，其目的是将工作区域、饭厅和休闲区变得像家一样。

2.2米X2米的豪华大床放在连墙的落地窗前，白色的床单和枕头营造出轻松的环境。在每个房间里，人们都可以抱着公仔Marvin（Gewoon设计）躺在床上欣赏LCD纯平电视播放的节目。床两侧的床头柜上摆放着泡泡纱的灯具。床下一巨大的暗柜可以放旅行箱和个人物品。

玻璃之内如同家一般
温馨溢满每个角落
这是奢华的气息

INTERIOR DESIGN
INSPIRATIONS

TOM FORD

HOW THEY WORK
AZ DESIGN
BRUCE WEBER
Caracas
Lima
Santiago
SUNDAY

流动艺术

Floating Art

汉堡机场拉迪森BLU 酒店
Radisson BLU Hotel, Hamburg Airport

设计公司：汉堡JOI-Design设计公司
项目地点：德国汉堡
项目面积：2935平方米
项目摄影：JOI-Design 设计公司

JOI-Design designed the interiors of a new-build Radisson BLU at Hamburg Airport. While the guestrooms follow the brand's prescription, the public areas depart into an innovative layering of oval forms within a large, open area where lighting and contrasting materials are used to "zone" space. The starting point for the interior design was derired from the constriction of the round architectural envelope. Acircular double height volume to the base of the building was topped by five three-quarter circles ofbedroom floors, projecting over the base and wrapping inwards on one side to create the main entrance from the terminal. All the public spaces – reception, lounge, bar, spa, fine dining restaurant, breakfast area and workspace – were to be accommodated within the round, ground floor.

A particularly intriguing feature is the door surrounds to two large wine chillers. With hundreds of inset wine bottle bases, back-lit to glow through the space, they are transformed into witty pieces of "glass-art" commissioned by JOI-Design from the acclaimed German glass sculptor, Sybille Homann. Throughout the public areas, hard finishes are juxtaposed with soft and fixed elements with floating, breaking down conventional barriers between work and leisure spaces while creating a sense off lexibility and mobility.

This is a modern business hotel, appropriate for its busy airport location, but it is also offers a number of intimate, even cosy areas where guests can work or dine, or simply unwind, at all times throughout the day.

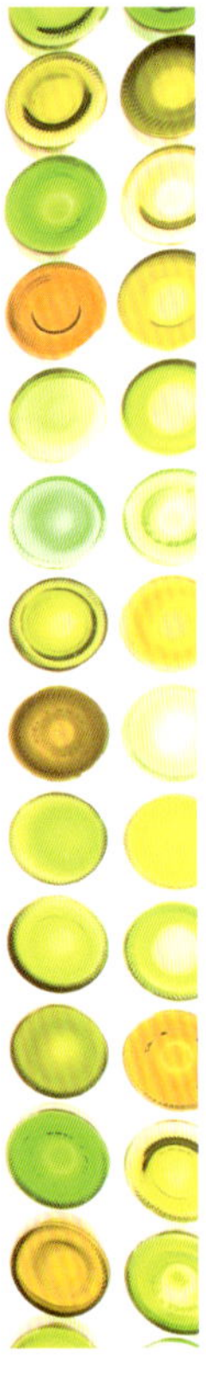

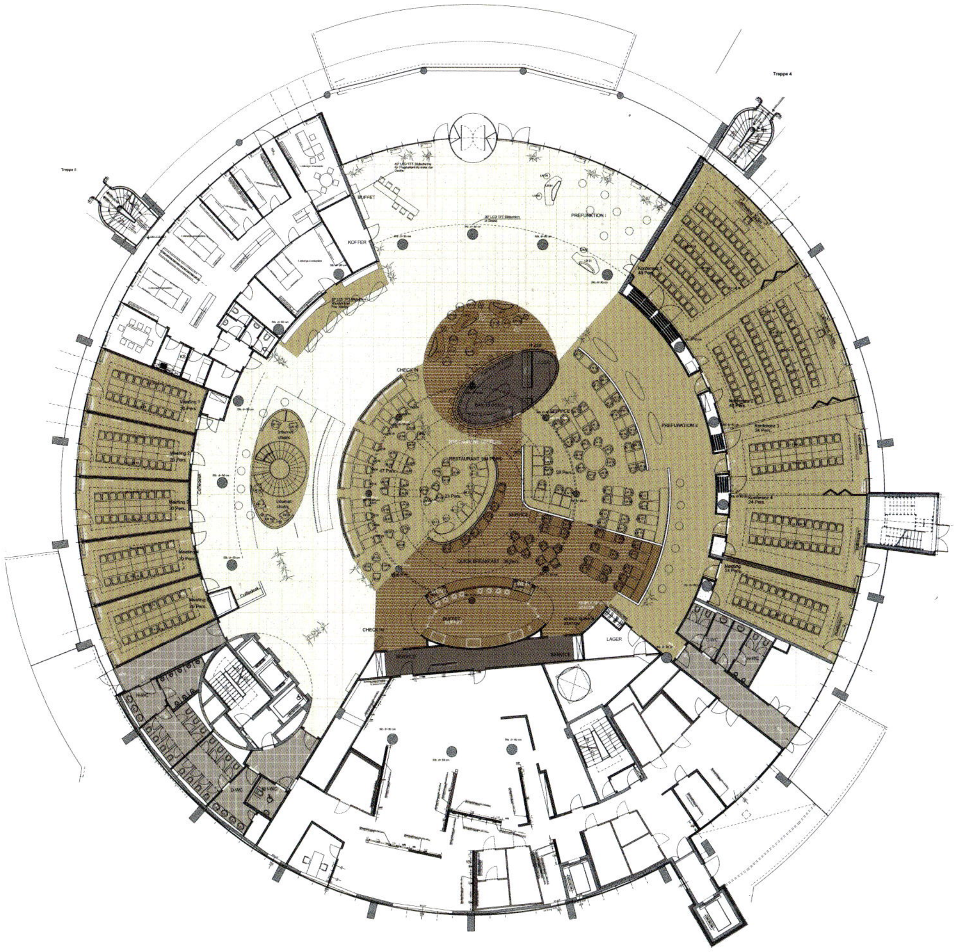

JOI设计公司完成了汉堡机场新拉迪森BLU酒店的内部设计。客房的设计遵循了该品牌酒店的传统，但在大气、通透的公共区域，设计师引入了一系列有机椭圆形，营造富有层次感的创新空间，并用灯光和对比材料“分隔”空间。内部设计的起点源于酒店建筑外壳的局限：基座是一个圆形的双层高空间，上面是5个3／4圆形的客房楼层，这5个楼层凸伸出基座，向内围绕基座的一侧，形成从机场出来后进入酒店的主入口。全部公共空间——前台、休闲吧、酒吧、水疗会馆、精美餐厅、早餐区和工作空间都容纳在这个圆形的一楼空间里。

一个特别引人入胜的特色是围绕着两个大型冷酒器的门。数百个装有背光照明的嵌入式酒瓶座让空间熠熠生辉，它们转化成了精巧的“玻璃艺术品”，这是JOI设计公司委托德国知名玻璃雕塑师Sybille Homann制作的。整个公共区域里，强硬的表面处理与柔和、固定的元素相辅相成，工作和休闲空间之间是打破传统的悬浮隔断，营造了灵活、流动的感觉。

这是一家现代化的商务酒店，适合其所处的繁忙机场位置，但也提供了若干温馨甚至惬意的地方，让客人们整天随时可以工作、进餐，或者什么都不做，只消尽情地放松休闲。

DAMPFBAD

玻璃艺术
强硬、柔和
相容又相隔

温柔的记忆
The Sweetest Memory

慕尼黑Dolce酒店与度假村集团
Dolce Hotels & Resorts Munich

设计公司：慕尼黑JOI-Design设计公司
项目地点：德国
项目面积：10 200平方米
项目摄影：慕尼黑JOI-Design设计公司

Soft materials and fine color palette constitute a comfortable environment. Walls are decorated in a minimalist way while the frames without pictures arouse people's imaginations. The materials inside the curved lobby are also mainly circles, achieving a perfect overall effect. Exclusive materials are employed to label a unique space. Oak tree, gilding, polished woolen goods, leather, concrete or cowhide become the symbols of different areas, which will be the sweetest memory for consumers in respect to the whole hotel. The designer tried to free the hotel from any sense of redundancy brought by previous hotel designers; instead, he preferred to express most people's desires to create a big place like a family. At the same time, every area is separated, but individual spaces are related with one another by the means of colors or materials.

舒适的环境
曲线的印记
带来温柔的记忆

柔软的材质和良好的色彩为人们提供了舒适的环境。墙体的装饰以简约为主，没有画面的镜框给了人们更多的想象，曲线造型的大堂内部的材质同样以圆形为主，构造一个良好的整体效果。本案利用专属的材料将空间设计成一个独特的标签。橡树、烫金、磨光羊毛织物、皮革、混凝土或牛皮成为空间不同区域的符号，对于整个酒店而言，这些符号能够成为消费者最温柔的记忆。设计努力摆脱了以前的酒店设计的冗赘感，设计师更加倾心表达大多数人的愿望，营造一处家庭式的大型场所。同时，每个区域之间是分开的，但是在独立的空间中也存在着关联，设计中采用的色彩甚至是材料将空间融合在一起。

光的欲望

The Desire of Light

亚特兰大W大酒店

W Grand Hotel of Atlanta

设计公司：Burdifilek

项目地点：亚特兰大

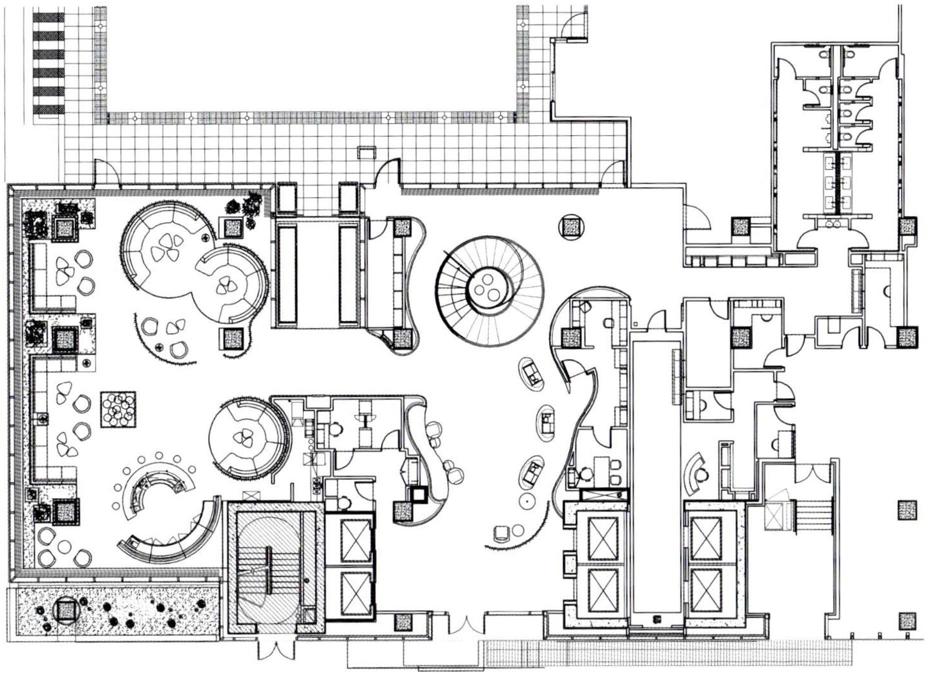

Burdifilek creates an urban oasis within Atlanta's newest W Hotel. With a few walks away from the downtown, W hotel is where natural elements meet the modern and contemporary living, and innovative sculptural objects were custom-designed by very talented designers all over the world, among which are renowned Canadian artists. Hidden treasures appear one by one while the guests move along the space as if it's a low-key art gallery. Hand-carved solid walnut wall with ripple surface constitute the background of the entrance and lobby. Above three reception desks, a glimmering mobile made of hundreds of metal leaves floats above the charcoal stone floor as it angles off delicate light, resembling a tree canopy. A custom-made metal sphere installation beneath a spectacular scooped staircase strengthens its theatrical atmosphere.

Burdifilek has used unconventional materials as major elements to bring us unlimited surprises. The bent-wood screens enclose an independent glamour area and curved seating perfectly matches with carpets, while polished stainless steel lanterns radiate a soft glow. A 22-foot water feature on the left-hand side becomes the space's background, which tranquilly trickles through metal lilies and fall vertically from the ceiling into a luminescent onyx reflecting pool.

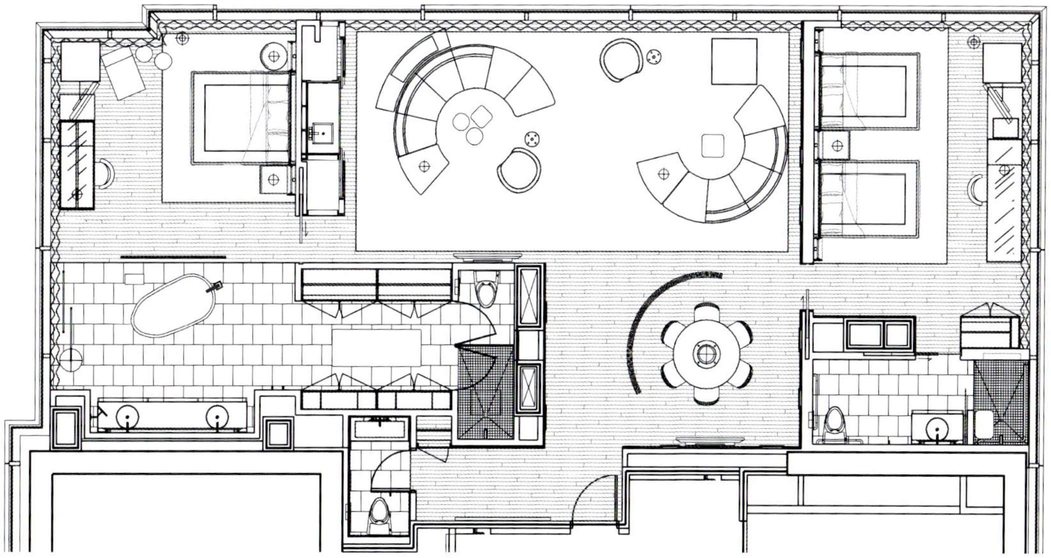

Burdifilek在美国亚特兰大的W大酒店创建了一处城市绿洲。W酒店距繁华的市中心仅几步之遥，自然元素在此融入到了现代都市生活中。这里的材质奢华，造型新颖，都是由全球最有才华的设计师定制的，其中包括许多知名的加拿大艺术家。

隐藏的珍宝随着宾客沿空间行进逐步展现出来，它像是一个低调的艺术馆。波纹状的手工雕刻胡桃实木墙形成入口和门厅的背景幕。在3个入住登记台的上方有数百片闪烁的金属薄箔片，它们悬浮在深灰色的石材地板上，折射出微妙的光，像是树的天蓬。在引人注目的勺形楼梯下面，一个定制的金属球体增强了这个展览馆的观赏性。

Burdifilek用非传统的材料构成主要元素，给人们带来无限惊喜。曲木屏风围成颇具魅力的单独区域，弧形的座位与地毯完美融合，抛光的不锈钢灯投射出柔和的光芒。左边一个0.56米长的水景成为此空间的背景，它沿着金属百合花茎的物体从天花板上垂直而下，潺潺地流向下方的黑玛瑙“水池”中。

RENAISS

这是城市的绿洲
那些散发着光芒的精灵
带着欲望温暖绿洲上的一切

爱琴海的诱惑
Temptation from Aegean Sea

瞭望台酒店
The Belvedere Hotel

设计公司：Rockwell Group
项目地点：希腊
项目摄影：Vagelis Paterakis、Ed Reeve

Inspired by the Aegean Sea, the project is a three-level space. The materials and design elements used echo those that are native to the sea. Guests enter on the top level, where is a bar that is comprised of a live edge wood bar top supported by a marble bar face. Hand-carved marble stones on stainless steel standoffs cover the bar face and are backlit to create the appearance that they are floating.

The restaurant and an additional bar are located on the lower level. The bar has a wood bar face that is carved to represent the movement of water. The walls of the main dining room are covered in white plaster. Carved-out wall niches are lined with mother-of-pearl inlay. The floor is made from stone with metal insets. Magdalena Globe lights, which are made of molded pieces of stamped metal, emit a glittering light that is similar ineffect to the sun seen underwater. The main restaurant on the lower level spills into an elegant pool area. This landscaped setting is designed to transform over the course of a day. During the daytime, there are single lounge chairs covered in terrycloth slipcovers with retractable sun shades. White matchstick screens shade single and double teak wood cabanas. In addition, we renovated one edge of the pool, adding new teak wood deck steps that skim the edge of the water.

流动的水和云
在爱琴海上嬉笑
快乐悠闲的人
在一方水域享受着自己的下午茶

灵感来自爱琴海，本案是一座三层酒店。采用的若干材料和设计元素均回应爱琴海的天然本色。客人从顶层进入，那里有一个酒吧，吧台用回收木材制成，台面是大理石。不锈钢支架上的手工雕刻的大理石覆盖着吧台表面，且有背光效果，制造了石雕仿如飘浮般的效果。

饭店和另外一个酒吧位于下一层。该酒吧的木质表面经过雕刻，形状象征着动态的水。主餐厅的墙壁用白色灰泥覆盖。墙上的壁龛镶衬着贝壳。地板用嵌着金属的石材制成。冲压成型的Magdalena球形灯发出一道道闪烁的光线，营造一种类似于阳光照射水面般波光粼粼的效果。

底层的主餐厅延伸到优雅的泳池区。这里的景色从早到晚不停地转变。白天，这里摆放若干罩着毛巾椅套、装着收缩式遮阳伞的单人躺椅。白色火柴棒式的屏风遮掩着柚木做成的单人和双人小屋。另外，设计师改造了泳池的一侧，增加了新柚木板台阶，水面掠过台阶，惬意感油然而生。

沐浴在经典

Be Immersed in the Classical

沐思城市酒店
Muse City Hotel

设 计 师：于丹鸿

项目摄影：施凯

供稿单位：与非视觉工作室

DiYuan Enterprise
帝苑企业
沐思城市酒店
MUSE City Hotel
九十八號
NO 98 RESTAURANT
新食坊
muse

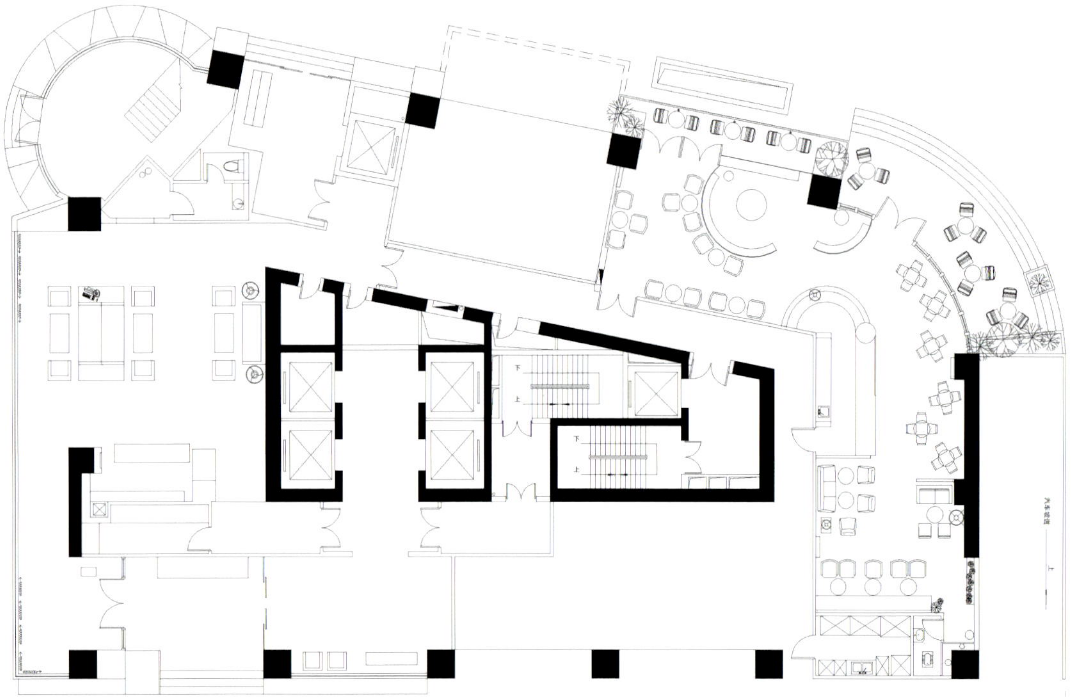

Situated in Ningde Building at the intersection of Wusi Road and Wenquan Road, the central business circle of Fuzhou, Muse City Hotel is a classic work that Diyuan Enterprise use to show its diversified brands. The hotel, focused on serving the groups in pursuit of living quality, is an agreeable habitat elaborately constructed out of the deep understanding of business and travel. Muse will interpret the definition of a city hotel from multi levels and create a quality living environment for people on business trip.

166 comfortable, pleasant rooms are created using 6 space designs different in style. The hotel insists on its own design ideas and fuses them into every detail the customers can feel. The hotel's design is not only about space layout, but more about interpreting Muse's attitude toward life from different angels to display its pursue for quality, bringing different feeling of business travel and brand new housing experience.

Putting an emphasis on a subtle spatial experience for the customers, the hotel started with every detail to create a spatial feeling suitable to most people. Interior space is not absolutely particular about luxury, instead, the designer inclined to more strengthen and deepen the intrinsic feel of the space itself. Simple modern design is used on guestrooms to present a fresh visual effect to the customers while rest areas is stressed with a comprehensive control of lightings and colors to provide a pure space for customers to have a good rest.

沐思城市酒店位于福州市五四路与温泉路交界处的宁德大厦，地处中心商务圈，是帝苑企业展示品牌多元化的经典力作。沐思城市酒店专注服务于追求生活品质的群体，出于对商务和旅行的深度理解而精心营建的宜居之道。沐思将从多个层面诠释城市酒店的定义，并营造出基于品质之上的商旅生活环境。

沐思城市酒店通过六种不同风格的空间设计而打造出166个舒适宜人的房间，酒店只坚持自己的设计理念，并将之融入顾客所能感受到的每个细节。沐思的设计不仅是对酒店空间的布局，更是通过不同角度来诠释沐思对生活的态度，从而体现其基于品质之上的追求，带来不同的商旅感受和崭新的居住体验。

酒店注重给予顾客细腻的空间体验，从每个细节上入手，营造可以适用于大多数人的空间感受。空间内部并不绝对讲究奢华的气质，设计师更加注重对空间本身所体现出来的感受的强化。客房采用了较为简单的现代设计，以清新的视觉效果呈现在顾客面前，而作为休息区，则注重对光线和色彩的综合把握，藉此提供给顾客绝对的休息空间。

沐浴着思念
在这里
寻找城市的经典

新概念
New Concept

福州市保罗假日酒店
Paul Holiday Hotel-Fuzhou

设 计 师：郑葳
设计公司：福州国广装饰设计工程有限公司
项目地址：中国福州
项目面积：500平方米
主要材料：镜面、玻璃、墙纸、地毯、仿古地砖、皮纹墙砖、软包、大理石等

Guestrooms, the main source of revenue for hotels, are used by the visitors for the most time after they check in, also the most intimate places. This case presents a new understanding of conventional decoration style for hotel guestrooms. Unlike the monotonously, conventionally designed guestrooms, it integrates a variety of elements. In view of limited usage area of the hotel space, an unlimited extension in space is visually created through a variation and imagination of design.

To remedy narrow, long corridors, large-scale tawny mirrors used on walls and ceilings not only improve the visual effect of the space but also bring out a feel of time tunnel. The walls of corridors are covered with wallpaper imitating woven straws and the arc patterns of rugs spread over the corridors result in a steady yet lively effect. The ceiling and door in the tone of white make the combination of black tawny mirror and quietly graceful wallpaper won't appear boring and depressing.

Because the targeted customers for this hotel are business travelers and the practical area of the space is limited, the design is required to be off the beaten track on the basis of continuing the essences of the convention. The dominant colors for guestrooms are relatively serious dark tones, simple but elegant colors. With the same approach as the corridors, mirrors, glasses and other diversified materials are used to soften too formal colors to avoid any depression, making the customers enjoy the warmth of a home once arriving inside the hotel as well as receive their visitors with decent courtesy. To satisfy the business hotel's diversified and varied needs, different types of guestroom are defined according to specific details under the premise of grasping the overall style. Each type of room not only satisfies the guest's different needs for a single room and double room, but also is designed with its own unique style.

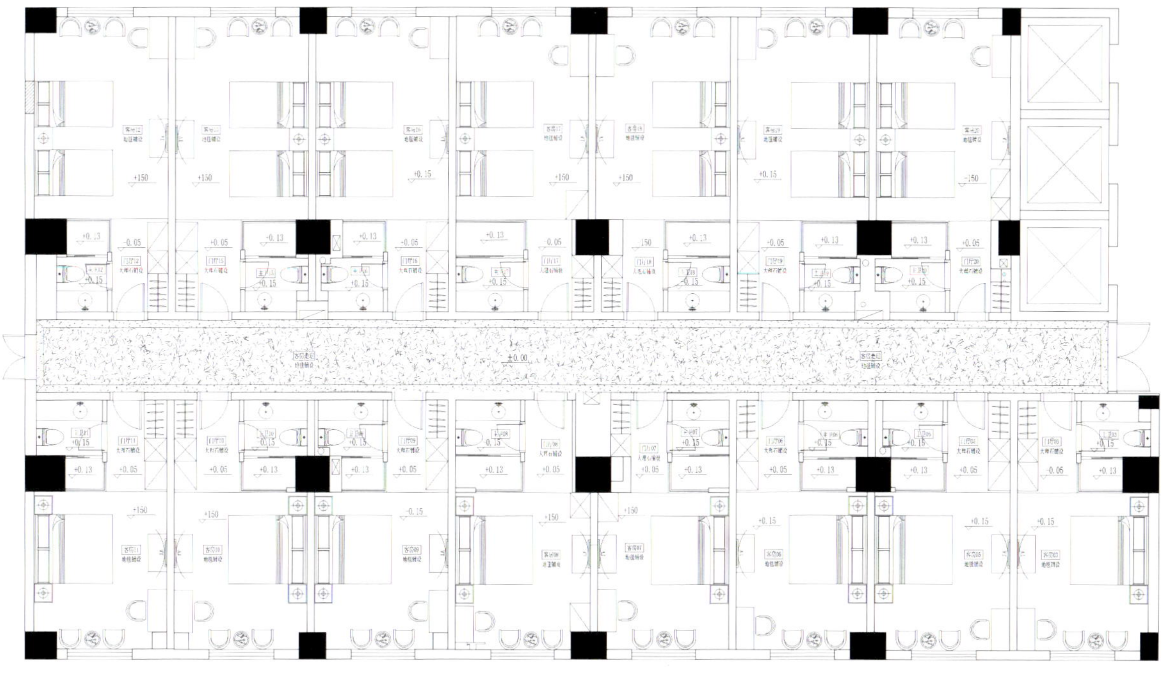

对传统的重新理解
赋予新的元素
让空间多一点想象

客房是酒店获取经营收入的主要来源，是客人入住后使用时间最长的，也是最具私密性的场所，而本案则是对传统酒店客房装修风格的重新诠释。它区别于传统设计式样单一的酒店客房，将多种元素融入其中。针对该酒店空间使用面积有限的情况，通过设计上的多变与想象，在视觉上创造一种空间上无限延伸的想法。

针对酒店走道狭长的空间，通过在墙面与层顶大范围使用黑茶镜，不仅改善了在空间视觉上的效果，而且营造出一种时空逶道的感觉。走道墙面采用仿草编墙纸，地毯的弧线图案使得走道在整体效果上稳重但又不失轻快。屋顶与房门采用白色作为基调，使得黑茶镜与素雅墙纸的结合不会显得过于沉闷。

由于该酒店对客源的定位在商务人士，且在空间上的使用面积有限，这就需要在延续传统精华的基础上又不落入俗套。客房在主体色调上采用较为稳重的深色调和素雅的颜色。与走道一脉相承的是，为了避免压抑，通过镜面、玻璃及其他多元化的材料来软化过于正式的色彩，使得顾客在入住时不仅有家的温馨，而且在有访客时又不失礼节。为满足该商务酒店多元化、多样性的需求，在把握大的风格前提下，根据具体细节制定出不同的房型。每个房型不但满足了客人对于单人房与双人房的不同需求，而且在设计上都别具独特风格。

东方神韵

Spirit of the Orient

鄂尔多斯瑞丽东方酒店

Ordos Rayli Oriental Hotel

设 计 师：翟勇、张迎军

设计公司：大石代设计咨询有限公司

项目地址：鄂尔多斯

项目面积：12 000平方米

主要材料：石材、不锈钢、水晶珠、木饰面

行李房
经理室
服务台
销售部
大堂吧
自助餐台
地下车库入口
信息查询
酒店大堂
±0.000
海鲜展示
服务总台
入口
餐厅
服务台
酒水库

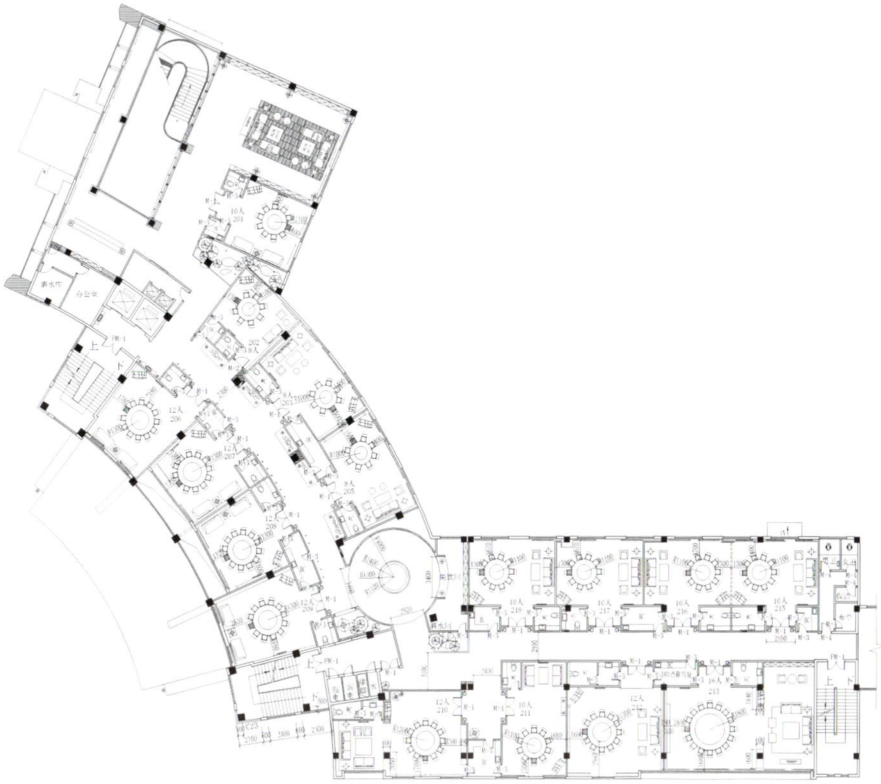

神秘的鄂尔多斯
在梦的前沿
热情而奔放的人们
浑身散发着东方的神韵

Situated in the richly endowed and mysterious city of Ordos, covering total 12,000 m^2, Rayli Oriental Hotel houses Delicacy Assembling Restaurant. The designers integrated charming oriental elements throughout the design echoing with the hotel's name, with oriental elements used as the striking key points of the theme, "oriental charms" are hidden in the minimalist design which also combines the local cultures–grassland, desert, national traits, etc., perfectly mixing the oriental philosophies with local cultures.

Crystal beaded curtains making up sophisticated floral patterns in the dinning hall reflect a perfect combination of modern materials and oriental elements. Custom-made rustic tiles covering the corridor walls highlight its Eastern charm. In the lobby, the service counter's backdrop is decorated with freehand ink brushwork, scholarly elegant without comprising its modern air. All works of art are carefully chosen by the designers to set off the incomparable atmosphere extremely and bring a top quality enjoyment to the customers.

Based on mainly grey, beige and coffee, the design tries to seek a subtle luxury and minimal magnificence through graceful, harmonious colors and oriental philosophies, elegant style and modern abstractness are gathered together to suit both refined and popular tastes. Reserved oriental cultures combined with post-modern colors give prominence to the eternal, everlasting oriental philosophies. In such a poetic space, a stunningly splendid art feast is presented to the visitors.

瑞丽东方酒店坐落于富饶而又神秘的鄂尔多斯市，总占地面积为12,000平方米，设有汇鲜坊餐厅。设计时融入了东方元素的神韵，这正与酒店名称相呼应，设计师将东方元素作为主题点睛，将“东方神韵”隐设在极简设计主题的背后，又结合了当地的地域文化——草原、沙漠、民族特征等，将东方哲学思想与当地文化完美融和。

餐饮大厅的水晶珠帘拼花造型精美地体现了时尚材质与东方元素的完美结合。走廊墙面为其量身定制的仿古砖，使酒店的东方韵味更加突出。酒店大厅服务台背景，也采用了大写意的笔墨字体装饰，儒雅而不失现代气息。酒店的所有艺术品均由设计师精挑细选，将酒店的气氛烘托得无以复加，为客人带来高品质享受。

酒店设计色彩以灰色、米色、咖啡色为主基调，意求低调的奢华、简约的华丽，色系效果素雅和谐，东方哲学、高雅风格、现代抽象，三者聚于其中，却能雅俗共赏。东方文化的内敛配合后现代的色彩搭配，突显了东方哲学的恒古不衰。在这样诗意般的空间中，为客人呈现了美轮美奂的艺术盛宴。

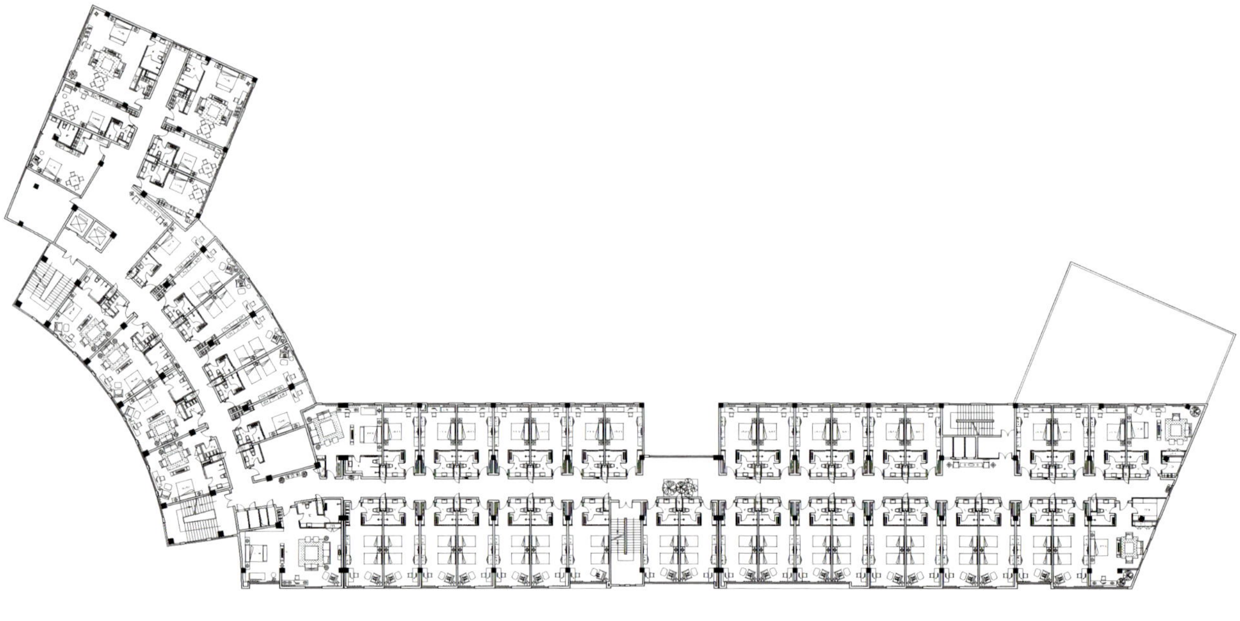

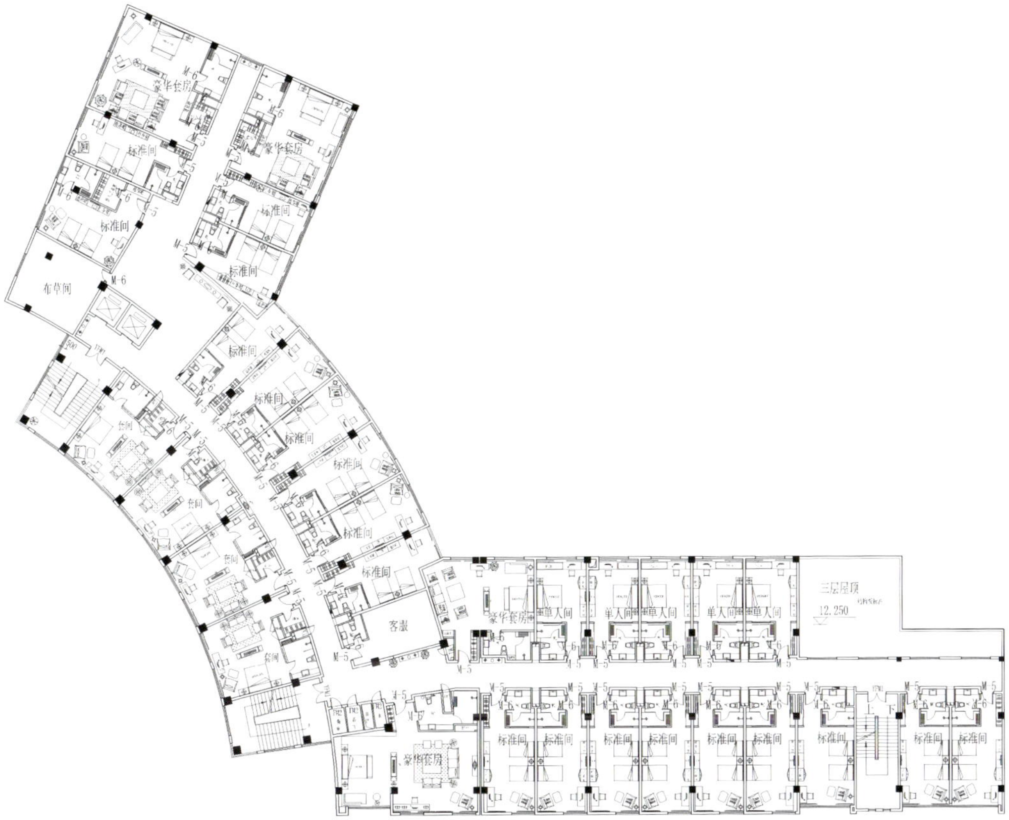

豪华套房
豪华套房
标准间
标准间
标准间
标准间
布草间
标准间
标准间
标准间
标准间
标准间
标准间
套间
套间
套间
套间
客梯
豪华套房
单人间
单人间
单人间
单人间
单人间
三层屋顶
12.250
标准间
豪华套房

6607

绿色风情

Green Flavor

迪拜国际沐浴主题酒店
Dubai International Bath Theme Hotel

设 计 师：李绍云、徐川

设计公司：恒通建设工程有限公司

项目地点：中国温州

项目面积：8400平方米

主要材料：家具板、米黄洞石、红镜、马赛克、天然透光云石

Dubai International Bath Theme Hotel, a super five-star green, leisure hotel built to international standards with huge investment from a Dubai company, integrates diversified functions like healthy bath, physiotherapy & health care, SPA, specialty food, multi-media lounge, large-scale shows, chess/card netbar, gym & recreation, children's playground and business guestrooms.

Covering a building area of over 10,000 m^2, it ingeniously elaborates a Mediterranean feel, where people are intoxicated with relaxing, happy bath while being exposed to soothing music and bathing fun, taste worldwide gastronomy in a cozy, elegant environment and experience an exotic air through its special SPA... All experiences of satisfaction and pleasure are merged here. The designers create a joyful city perfectly combining the luxury space with refined details in an easy yet discreet manner. Reasonable layout, elegant yet intimate VIP areas offer a socialization platform with high taste, signifying unlimited business opportunities.

From the very beginning, the designers decided to introduce a Mediterranean feel into the space and the subtle techniques can be seen everywhere, for example, the designers chose to lay an emphasis on colors and lightings to build up the space, in which warm, graceful soft yellow lends a relaxed, comfortable feeling, contrasted bright and dark lightings increase the sense of tiers to make customers feel the strong temperament transmitted from the space. Moreover, the designers took advantage of main interior structure to give an absolute exotic flavor to the space.

8035
8039
8051
8036
8038
8050

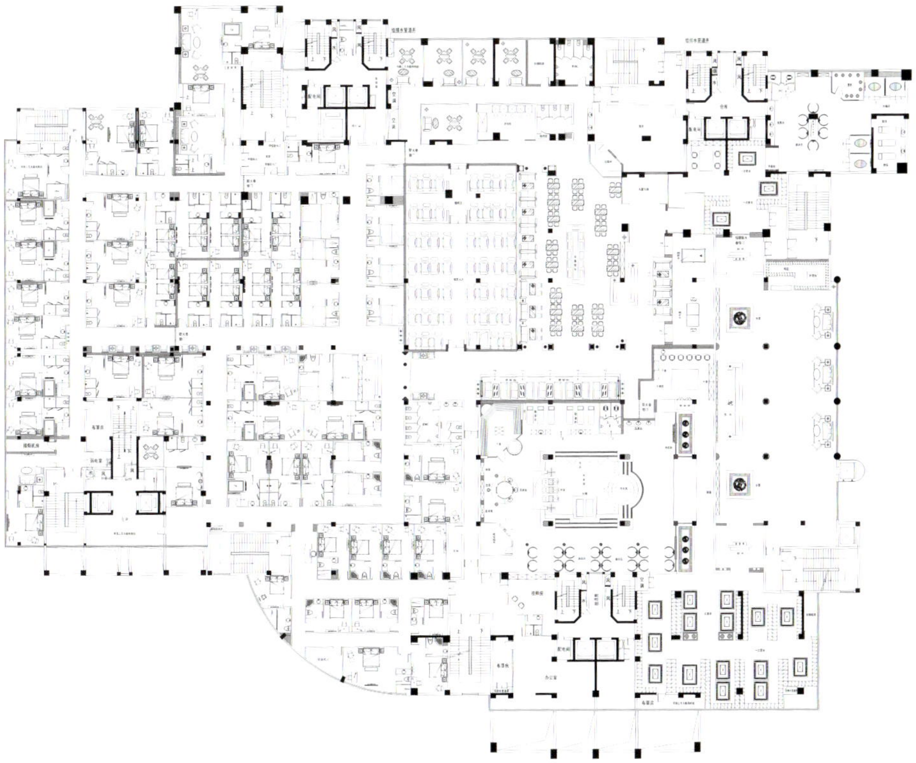

迪拜国际沐浴主题酒店是由迪拜公司斥巨资按照国际化标准兴建的超五星级大型绿色休闲场所，集健康沐浴、理疗保健、SPA水疗、特色餐饮、多媒体休息厅、大型演艺、棋牌网吧、健身游艺、儿童乐园、商务客房为一体。建筑面积10 000余平方米，独具匠心地演绎地中海风情，沐歌浴趣中醉享身心俱舒的洗浴欢乐，舒适之余品味世界美食，特色SPA中感受异域风情……所有的享受体验在这里交融汇集，设计师用轻松而又不失尊贵的手笔，营造了一个奢华空间与精致细节完美融合的欢乐之城。合理的空间布局、优雅而又私密性绝佳的尊贵VIP专区，更是为您提供了一个商机无限的高品位交际平台。

设计从最初就决定将空间引入地中海风情，因而在设计的过程中，细腻的手法处处可见，设计师选择在色彩和光线上着力打造空间，具有温暖淡雅感受的暖黄色营造出了空间简约、舒适的氛围。光线的明暗对比处理，增添了空间的层次感受，让顾客可以在这里强烈感受到空间所散发出来的气息。此外，设计中利用室内主体结构，让空间获得绝对的异域风情。

地中海风情
沐歌浴趣
这是欢乐之城

风景无限好

Infinite Beautiful Scenery

海源国际大酒店
Haiyuan International Grand Hotel

设 计 师：郭宇宏、吴振华、郑岩川、杨茂果、徐朝霞、林振伟

设计公司：福州佐泽装饰工程有限公司

项目地点：中国福建

项目面积：30 500平方米

HAIYUAN HOTEL

凭窗望
好风光
梦里几缕斜阳

Located in Putian city, adjacent to Puxian Theater, Haiyuan International Grand Hotel enjoys incomparable geographic advantages, overlooking fabulous scenery. It's only 5 minutes drive from downtown Putian and 10 minutes drive from the railway station. With the total building area of 30,500 square meters, this is a five star business hotel boasting 208 guestrooms up to international standards, ranging from 22.5 to 135 square meters per room (suite). Standard suites, sophisticated luxury suites and president suites continue the good reputation of the spacious, comfortable hotel, where people can lean against a window and look into the distance to enjoy panoramic, gorgeous views. All guestrooms are equipped with full set of amenities, including broadband internet access, data interface, international direct dialing, satellite TV, utilities for preparing tea/coffee and 24-hour room service. 660 m^2 western restaurant can accommodate 160 people to dine together, 334 m^2 lounge can hold 85 people, big and small banquet hall with the respect area of 444.7 and 280 m^2 are able to entertain 196 and 160 while multifunctional hall of 655 m^2 is capable of a banquet and conference for 255, besides, there're 21 individually styled private dinning rooms sized between 32 m^2 to 68 m^2, which all have their own characteristics. The area of lobby is 675 m^2. Moreover, the hotel contains complete fitness and recreational facilities, among which are 19 single to four-people sauna rooms ranging from 25 to 135 m^2. For KTV, there're 20 diversified private rooms spanning from 35 to 73 m^2.

海源国际大酒店地处莆田市，毗邻莆仙大剧院而建，俯瞰迷人风光，地理位置无与伦比。距离莆田市中心只需5分钟车程，离火车站也只有10分钟的车程。酒店建筑总面积为30 500平方米，是一家五星级商务酒店，拥有208间（套）22.5 平方米到135平方米的国际标准客房、标准套房、各式高级豪华套房和总统套房，构造延续了酒店宽敞舒适的美誉，凭窗远眺，美丽风光尽收眼底。

所有客房都提供全套便利服务，包括宽带上网、数据接口、国际长途直拨、卫星电视、茶水、咖啡、冲泡设备，以及24小时客房服务。660平方米的西餐厅可容纳160人同时就餐；大堂吧344平方米，可容纳85人；大小宴会厅分别是444.7平方米与280平方米，同时分别可容纳196人与160人就餐；多功能厅655平方米，可容纳255人参加宴会与会议；以及各式风格迥异的餐厅包间21间，面积在32平方米至68平方米不等。大堂面积675平方米。此外，酒店还设有齐全的康乐设施和配套服务设施。其中，桑拿拥有19间面积为25平方米至135平方米的单人至4人包间；KTV拥有面积35平方米至73平方米不等的20间各式包间。

沙漠装饰
Desert Deco

希尔顿度假俱乐部大酒店
Hilton Grand Vacations Club

设计公司：Watg设计公司
项目地点：美国拉斯维加斯

Create a lightly themed, vacation ownership property in Las Vegas. Being the third vacation ownership property owned by Hilton in Las Vegas, it needed to be its own distinct design to attract new and repeat timeshare owners desiring to vacation on the exciting Las Vegas Strip.

The project design was christened "Desert Deco", in celebration of the 20th Century Art Deco era and its influence in the Desert Southwest. The three story atrium lobby is a dramatic entrance which sets the tone for the project's design. The guestroom units are colorful and contemporary with many desirable amenities and an impressive view of all that Las Vegas has to offer. There is a well appointed, state of the art Sales Center, which provides a top notch sales tool for potential timeshare owners.

在拉斯维加斯创造一个主题轻松的分时度假产权酒店。作为希尔顿集团在拉斯维加斯拥有的第三个分时度假房产项目，它需要以自身独特的设计来吸引新客及回头客，这些客人渴望以分时享用酒店所有权的方式在令人兴奋的赌城大道度假。

项目设计命名为"沙漠装饰"，以此纪念20世纪的装饰艺术时代及反映这种风格在内达华州西南沙漠区的影响。3层高的中庭大堂构成令人惊叹的宏伟入口，奠定了项目的设计基调。客房单元五彩缤纷，现代气息浓厚，有许多令人称羡的设施并坐拥拉斯维加斯最好的壮观风景。设备完善、技术先进的售楼中心为潜在的分时用房者提供了一流的销售场所。

Hilton
Grand Vacations Club

五彩缤纷之间
沙发的装饰
是对过去的纪念

高雅的体验
Elegant Experience

海泽顿酒店
Hazelton Hotel

设计公司：Yabu Pushelberg设计公司
项目地点：加拿大多伦多
项目面积：15 236平方米
项目摄影：Evan Dion

The hotel imparts the warm intimacy of a manor home. Classic designs receive a modern update with inviting textures and superb craftsmanship. Throughout the public spaces, the handsome tradition of a men's club is graced with feminine touches such as mother of pearl insets and mirrored bi-fold screens.Designed in the flavour of a large great room, the lobby space is defined by decorative metal screens which add artistic interest. Walls are covered in supple leather panels, while a custom cove ceiling features a sculptural scallop pattern. Dressed in a subdued palette, the lobby has a masculine flavour which sets the tone for the hotel and is interspersed with moments of glamour.

Located off the lobby is a small jewel box of a reception area. Paneled in charcoal grey cowhide, the space has a refined sophistication. A sculptural bronze reception desk makes a luxe statement.

Guest rooms are adorned in rich textures and a warm palette with all of the modern conveniences of a five-star hotel. Custom leather headboards wrap around the beds in an elegant L-shape. Select furniture pieces lend a residential in feel to the rooms. Large walk-in closets in wood and granite provide a well-appointed dressing room. Luxurious bathrooms boast floor-to-ceiling granite in a rich charcoal stone with a large bathtub, shower and spacious double vanity. Three presidential suites each include a round rotunda. Intended to impart a welcoming residential quality, offering an elevated experience.

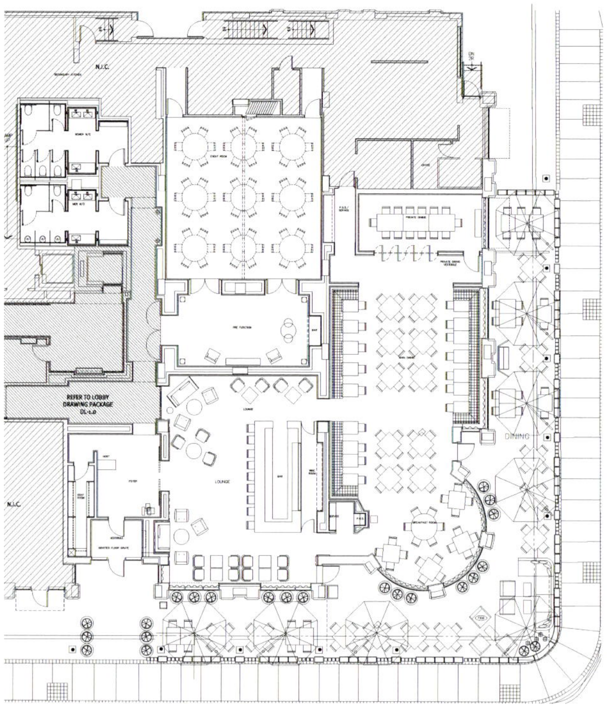
REFER TO LOBBY
DRAWING PACKAGE
DL-1.0
N.I.C.
LOUNGE

酒店散发着庄园般的温馨和亲切感。空间注重材质和精湛工艺的运用，古典设计带给人现代之感。在整个公共空间中，一个有着男人潇洒气质的俱乐部却又能给人以女性化的触动。本着设计一个大空间的初衷，设计师在大堂空间中运用了装饰金属屏幕，从而使空间充满了艺术感。墙壁以柔软的皮革面板为装饰，特制的拱形天花板则雕刻着扇贝图案。整个酒店浸润在柔和的色调之中，阳刚味十足的大堂为酒店定下了基调并增加了整个酒店的魅力。

在大厅外是一个包厢似的接待区，经过炭灰色的皮质打造，整个空间洋溢着一种精致的设计感，雕琢的古铜色的接待桌使得整个设计变得更加华丽。

客房用华丽的材质打造出一个五星级现代的具有暖色调气质的酒店。订制的床头板以L型的样式包裹着客床。精选的家具为房间增添了几分居家的感觉。木质与花岗岩的大型衣橱构成一间设备齐全的更衣室。豪华的浴室铺满了花岗岩与木炭石，内有巨大的浴缸与淋浴器以及双洗脸盆。3 间总统套间各带一个圆形大厅，宾至如归的居家感受营造出一种高雅的体验。

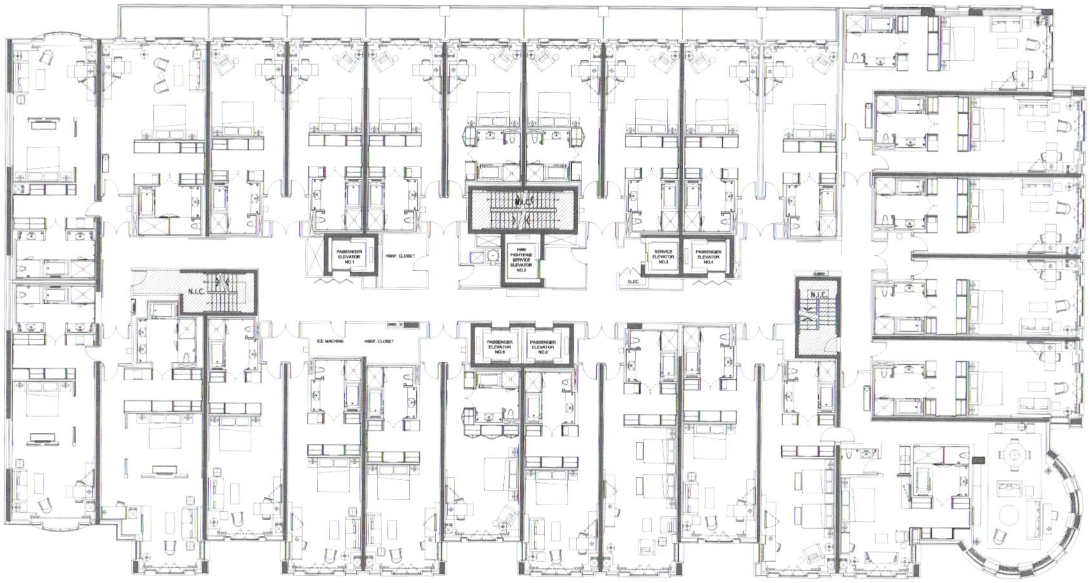

呼吸庄园的清新
古典中孕育现代的胚胎
是男人潇洒的气质
是女人温柔的触动

墨西哥艺术

Mexican Art

Las Alcobas——海兹顿酒店

Las Alcobas - Hazelton Hotel

设计公司：Yabu Pushelberg 设计公司

项目地点：墨西哥城

This seven-story small luxury hotel Las Alcobas was designed to reflect the property's namesake (Las Alcobas translates to "the alcoves" in Spanish) and aptly expresses the intimate and modern guest house concept. Located in Polanco, one of Mexico City's most fashionable districts, the hotel offers 36 guests rooms and suites including three penthouses with wraparound terraces overlooking the city. In addition to designing the exterior of the fully remodeled and modernized 1950s building, Yabu Pushelberg has created a modern interior that is infused with warmth and texture, reflected in the custom-made furnishings, hand-knotted rugs and original artwork by contemporary Mexican artists that nourish the senses throughout. Much like the spiral wood and stone staircase that serves as the focal point of the hotel, every architectural detail and fine-finish has been considered, from stunning full slabs of light blue marble, stitched-leather wall panels and inlaid wall carvings to stone finishings and the use of richly-colored woods throughout.

温馨而时尚的墨西哥艺术
典雅而精致
阳光带来温馨的祝福

该7层高的小型豪华酒店Las Alcobas是以业主的名字命名的（Las Alcobas是“the alcoves”的西班牙语翻译），确切地诠释了温馨而现代的客房设计理念。该酒店坐落在墨西哥城最时尚的地区——Polanco。36个客房中的3间楼顶房被露台包围，从这里可以俯瞰城市美景。除了对20世纪50年代的这座建筑彻底翻新，并使其具有现代风貌外，Yabu Pushelberg还设计出一个温暖而富有纹理的现代室内环境，订制的家具和手织地毯充分体现出这一点。现代墨西哥艺术家的原创艺术作品更是滋养着这丝意味。螺旋纹的木板和石制的楼梯成为了酒店的焦点，每一个建筑的细节和精加工环节都被考虑到了——从浅蓝色大理石上令人惊艳的厚板、刻有刺绣的壁板、嵌入式的墙面，到石制硬装和色彩艳丽的木质面。

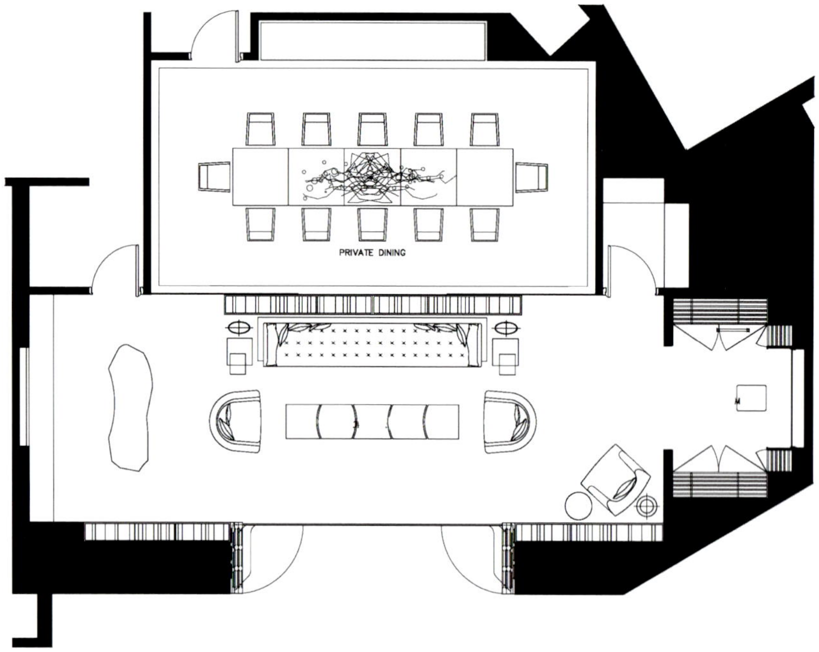

表情
Expressions

MINE精品酒店
MINE HOTEL BOUTIQUE

设 计 师：Adriana Sternberg、Mariano Kohen、
Camila Chulak、 Lucía De Chiara
设计公司：Sternberg Kohen Arquitectos设计事务所
项目面积：932平方米

The program consists of a ground floor containing the hotel's services and public spaces: the lobby, the central patio, the lounge, the restaurant, the garden and swimming pool. These are organised as a sequence of interior and exterior spaces linked by the circulation, (in some cases defined by their borders, in others virtually), that goes through them with the intention of making the observer lose sense of whether he/she is inside or outside the building.

The twenty rooms are organised in the three superior floors, forming two blocks linked by the central patio, which now becomes the core of the project and contains the vertical and horizontal circulations. The first block of rooms faces the street and the central patio; the other block faces the garden.

The project is organised around two ideas; in the first place, the "proménade"; unlike most chain hotels, circulations are not dark, residual spaces, but a sequence of different looks of the building and its surroundings, discovering the block's interior and the other gardens in Palermo.

The second idea in which we worked on was to get a building expression that could, on one side, express the architecture of this kind of hotels: clean lines, net spaces and special attention to the details; and, on the other side, that could merge with its surroundings. This is what helped us define the fusion between shapes and materials. Exposed concrete, micro cement and glass coexist with leather, stone, textured walls and earth colors. The façade's result was an accurate expression of such fusion.

设计方案包括位于一楼的酒店服务区和公共空间——大堂、中庭、休闲吧、餐厅、花园和泳池。这些地方组成井然有序、相互连接的内外空间（有些区域按其边界来界定，另一些则是很直观的一目了然），让穿行其中的人们不知他（她）究竟是在建筑里面还是建筑外面。

20间客房安排在3个楼层，组成由中庭连接的两座楼，现在这里成为项目的核心，包含了竖向和横向通道。第一座的房间均朝向街面和中庭，其他房间朝向花园。

项目围绕两个构思，首先“通道”与大多数连锁酒店不同，这里的通道不是阴暗的剩余空间，而是通过它们看到酒店及其周边的不同面貌，发现楼座的内部及巴勒莫的其他花园。

设计师的第二个构思是获取某种建筑表情，一方面表达出这种酒店的建筑特征：简洁的线条、实用空间和对细节的特别关注；另一面，能够与周围环境融合。这些都有助于让我们确定形状和材料之间的结合点。暴露的混凝土、超细水泥和玻璃、皮革、石材，肌理墙面和大地色融合为一体。立面的处理效果正是这种融合的表现。

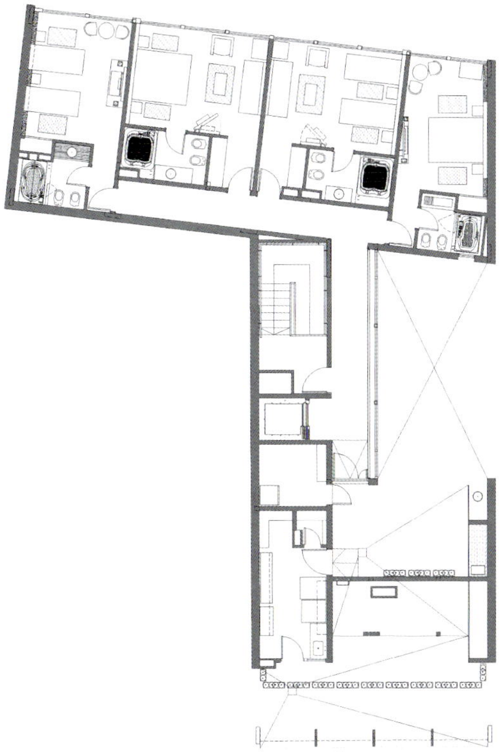
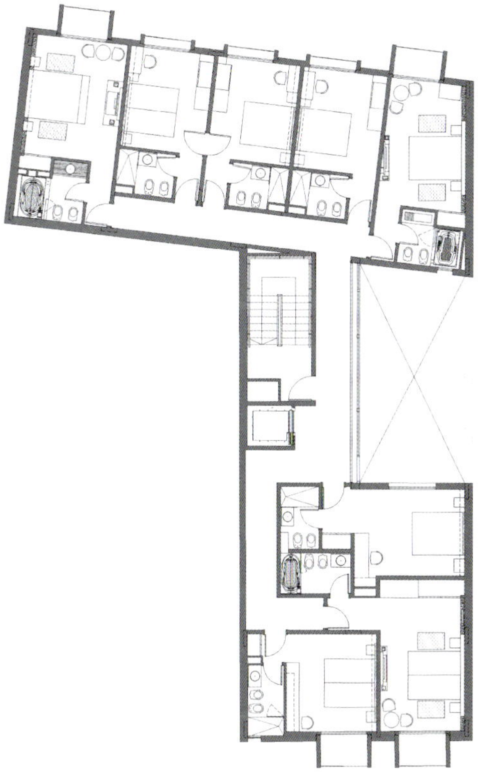

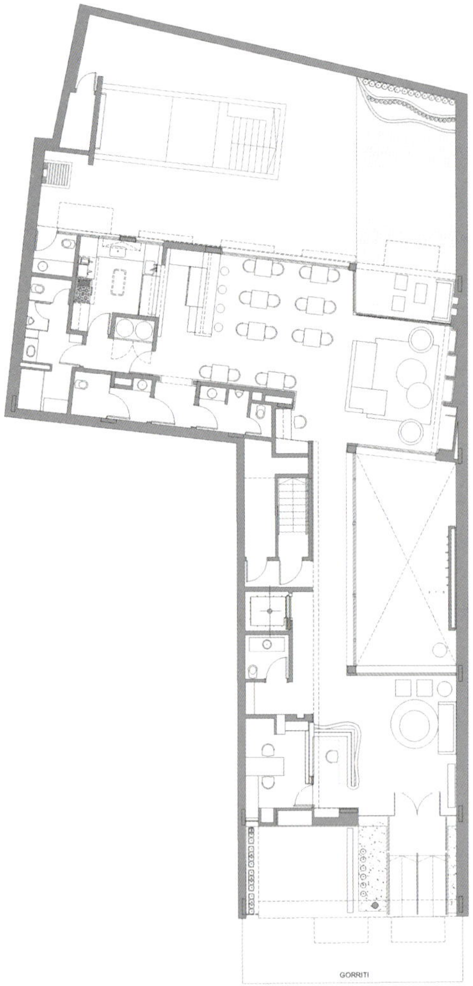
GORRITI

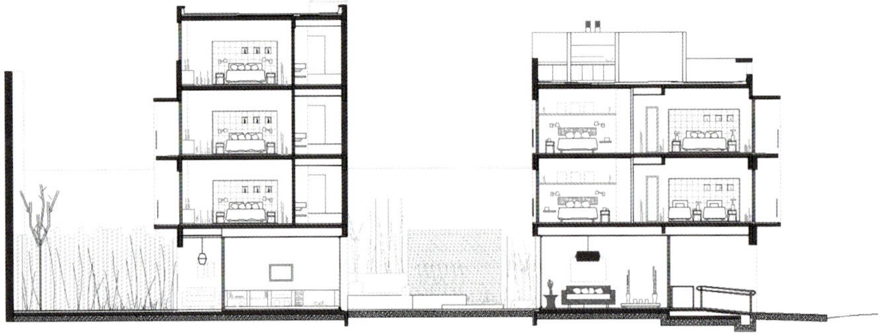

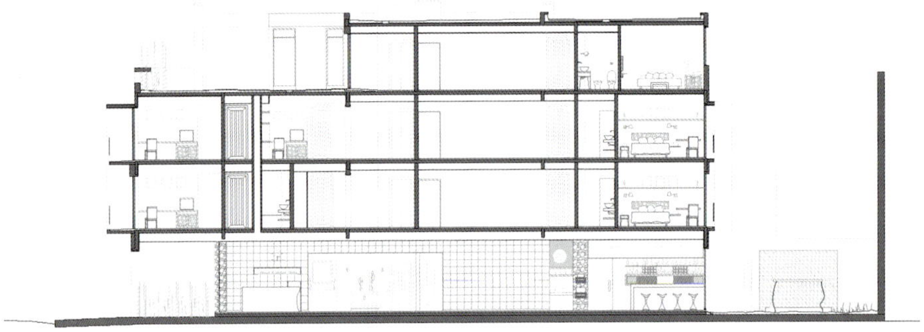

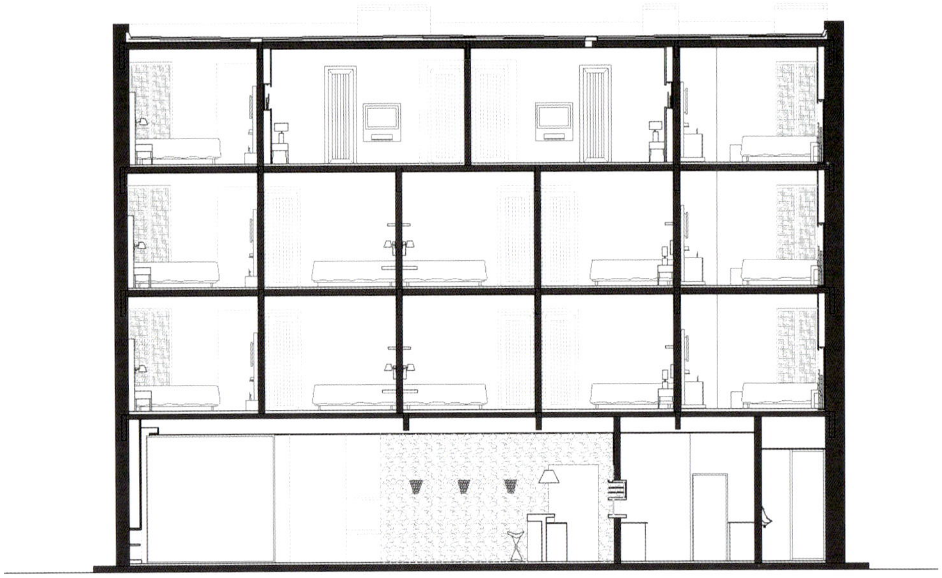

建筑的表情
是细节的呼吸
各方的使者
在这里朝会

TOP TEN CHINESE DESIGNERS 中国十大当红设计师

邱德光

邱德光设计事务所主持人暨总设计师
被誉为新装饰主义大师
中国设计界的领军人物

邱德光（T.K.Chu）30年来致力于两岸的尖端室内设计，以丰富的经验与深厚的素养，将装饰元素结合当代设计，开创了新装饰主义（NEO-ART DECO）东方美学风格，熟悉运用东方华丽、艺术、时尚元素，将生活形态和美学意识转化为尊贵身份，赋予奢华生活新内涵，成功塑造当代东方都会美学与21世纪时尚多元的生活形态。

TOP TEN CHINESE DESIGNERS 中国十大当红设计师

孙华锋

孙华锋，上海理工美联大学室内设计硕士研究生；高级室内建筑师。河南鼎合建筑装饰设计工程有限公司总经理。现任中国建筑学会室内设计分会常务理事、中国建筑学会室内设计分会（第十五）专业委员会主任、《INTERIOR DESIGN CHINA》杂志郑州联络处主任等职务。

近年获奖情况:
2010年《室内公共空间》杂志社“2010－2011中国十大当红设计师”
2009年CIID中国室内设计二十年“杰出设计师”
2009年中国室内设计大奖赛商业工程类二等奖
2009年第十七届亚太地区室内设计大奖赛荣誉奖
2009年中国风-IAI 2009亚太地区室内设计邀请赛铜奖
2008年美国INTREIOR DESIGN年度十大封面人物
2008年亚太地区室内设计双年大奖赛铜奖
2008年中国室内设计大奖赛酒店、宾馆工程类三等奖
2008年上海“金外滩”最佳色彩运用奖
2007年IFI第三届国际室内设计大奖赛优秀奖及佳作奖
2006年IFI第二届国际室内设计大奖赛商业类三等奖
2005年中国室内设计十大年度人物提名奖

TOP TEN CHINESE DESIGNERS 中国十大当红设计师

林文格

林文格，L&A文格空间设计顾问(深圳)公司创始人，高级室内建筑师、国家杰出室内建筑师。现任IFDA国际室内装饰设计协会理事、ICAD国际A级职业景观设计师、中国建筑学会室内设计分会（第三）专业委员会委员、香港室内设计协会中国深圳代表处委员、中国建筑学会会刊《a+a》顾问委员会委员。意大利米兰理工设计学院室内设计硕士。

近年获奖情况：

2010年《室内公共空间》杂志社“2010—2011中国十大当红设计师”

2009年美国Hospitality Design Awards酒店空间设计大赛最高荣誉winner大奖

2009年世界酒店“五洲钻石奖”——“最佳设计师”

APIDA第十一届亚太地区室内设计大奖餐馆酒吧类别冠军奖

APIDA第十四届亚太地区室内设计大奖酒店类别铜奖

2008年“华耐杯”商业方案类二等奖

2007年“金外滩”提名奖

2007年“金外滩”最佳餐饮空间奖

2006年INTERIOR DESIGN 酒店设计奖

2006年中国（深圳）室内设计文化节城市荣誉奖

2006年中国（深圳）室内设计文化节酒店设计类别金奖

2006年第三届海峡两岸四地室内设计大奖赛五项大奖

2006年《现代装饰》（国际）室内设计传媒奖年度精英设师

2006年中国建筑学会“全国百名优秀室内建筑师”

2004年中国建筑学会“全国百名优秀室内建筑师”